PERFORMANCE OF EUROPEAN WIND TURBINES

A Statistical Evaluation from the
European Wind Turbine Database EUROWIN

A report based on work done by the Fraunhofer Institute, Freiburg, under contracts from the Commission of the European Communities, Directorate-General Science, Research and Development (DG XII), as part of the European Community's Non-Nuclear Energy programmes from 1985–1992.

PERFORMANCE OF EUROPEAN WIND TURBINES

A Statistical Evaluation from the European Wind Turbine Database EUROWIN

J. SCHMID

and

H. P. KLEIN

*Fraunhofer Institute for Solar Energy Systems,
Freiburg, Federal Republic of Germany*

ELSEVIER APPLIED SCIENCE
LONDON and NEW YORK

ELSEVIER SCIENCE PUBLISHERS LTD
Crown House, Linton Road, Barking, Essex IG11 8JU, England

Sole Distributor in the USA and Canada
ELSEVIER SCIENCE PUBLISHING CO., INC.
655 Avenue of the Americas, New York, NY 10010, USA

WITH 23 TABLES AND 90 ILLUSTRATIONS

British Library Cataloguing in Publication Data
Schmid, J. (Jürgen)
  Performance of European wind turbines.
  I. Title   II. Klein, H. P.
  621.312136

ISBN 1-85166-737-7

Library of Congress CIP data applied for

Publication arrangements by Commission of the European Communities, Directorate-General Telecommunications, Information Industries and Innovation, Scientific and Technical Communication Unit, Luxembourg

EUR 13929 EN

Printed in Great Britain at The Alden Press, Oxford

Preface

The Commission of the European Communities supports a large range of Research and Technical Development programmes whose value lies not merely in the financial contribution to activities that might otherwise not take place but in the opportunities which arise to establish truly Europe-wide networks of collaborators.

Neither these nor any other researchers are, however, complete, or indeed have any value, until their results are reported. Neither is this the end of the matter because it is also necessary that those who may potentially benefit from the work are aware of its existence and can consult and utilise the findings. The CEC has always recognized the importance of disseminating the results of the work which it funds and by doing so helps to further the sense of truly pan-European scientific Communities in the technical areas which it supports.

Wind Energy R&D has formed part of the multiannual CEC non-nuclear energy programmes since 1981. Its results have been widely disseminated through Conferences, Workshops, papers and other such traditional routes. In addition, however, it was realised that the programme which has recently been completed would be appropriately reported in a series of books summarising the results of one or several related activities; this would be particularly valuable to those who may want to understand the outcome without delving into the technical minutiae.

This is the first of the series which describes the results of a single, but most important, activity. With the rapid growth of wind energy application for grid connected systems in Europe towards the end of the 1980s, a large amount of experience related to the technical and economic aspects of installation and operation is beeing gathered. The advantages of these data and interpreting them are manifold. Dr.J. Schmid and Mr.H.P. Klein at the Fraunhofer-Institute in Freiburg have worked on their EUROWIN database for a number of years and have now wealth of experience as the database grows through the addition of new wind turbines and new facets of behavior.

This book summarises these results and draws out much information which is of importance to researchers, manufacturers and utilities. I thank the authors for their excellent contribution, as well as the contacts in many EC countries who have supplied data and continue to do so. I commend the results of their work which give an excellent start to this series and well fulfills the twin objectives of carrying out good quality research and reporting the results in an interesting and stimulating manner.

Wolfgang Palz
CEC-DG XII
Bruxelles

Acknowledgements

This book was made possible by an international collaboration of wind energy experts from eight member countries of the European Community. We wish to thank all the sub-contractors for the hard work they have done in gathering the operational data which are the basis of the book.

Special thanks go to Mr. Helmut Godard of ENERGOSSA GmbH/Freiburg who did a large part of the programming work in the last phase of the project.

We are greatly indebted to Dr.Richard Shock, expert to the Commission of the European Communities DG XII, and Mr.David Milborrow of National Power/UK for their scientific and editorial input. Many ideas, especially for graphic presentations, were also contributed by these two colleagues.

We are very grateful to Laurence Crossby/UK and his wife Jean for the two days H.P. Klein spent with them in Sussex/UK for the final editorial work. L. Crossby shared the difficult task of translating our 'broken English' into acceptable and understandable sentences.

Furthermore we would like to thank Anne Kovach from Berkeley/Ca./USA who helped us in the initial stages of writing.

We thank the Commission of the European Communities DG XII and the Wind R&D - programme manager Dr.Giancarlo Caratti for the financial support and guidance.

Very special thanks go to Dr.Wolfgang Palz, head of the R&D - programmes in Renewable Energies for his continuous support, motivation and his numerous contributions to this project.

Freiburg, July 1991   Jürgen Schmid

Last, not least, 'muchas gracias' to my wife Guadalupe, to whose questions about the progress of the book and when it would be finished I always replied 'Mañana'.

Biltzheim, July 1991   Hans-Peter Klein

CONTENTS	page

Executive Summary

"Performance of European Wind Turbines" is the result of a comprehensive collaboration between most of the European member countries which has been initiated and financially supported by the Commission of the European Communities DG XII. The project started in August 1985 in the framework of the second CEC DG XII Wind Energy R&D - programme and the results presented here can be regarded as representative of the state of the art in European wind energy technology.
The statistical analyses are based mainly on the monthly inflow of operational data of more than 3500 plants (status July 1991). These comprise individual wind turbines as well as windparks. Each windpark is treated as one plant in the European wind turbine database EUROWIN. The EUROWIN database can be accessed by researchers, designers and those, active or interested in wind energy research. The collection of data is based on questionnaires with the exception of Denmark and The Netherlands for which data are transferred directly from the national databases. The database consists out of sub-files in which not only operational data are stored, but also technical details of the different wind turbine types (component weights, rotor blade data, transmission, generator, failure data, component costs, etc.).

Based on this information, the following types of output have been generated :
- total number of turbines, installed capacity
- specific power
- design wind speeds
- component weights
- mean turbine size

as examples of technology analysis.

The statistical analyses of operational data provide information on individual and mean energy production, capacity factor, operational time, failure analysis, economics, and energy pay back time.

This information is provided for specific time priods, countries, sizes, and technology concepts.

Some of the most important results are :
1.   The number of installed windturbines as well as wind capacity is increasing rapidly.

2.   The mean size of wind turbines is continuously increasing, from 80 kW in 1986 to 160 kW in 1989.

3.   The mean specific energy output is continuously increasing, the best machines improved from 1000 [kWh/m² a] in 1986 to nearly 1600 [kWh/m² a] in 1989.

As a consequence, the economics of electricity production by means of wind turbines has been improved continuously and costs on good sites are now competitive, and in some cases lower than those of certain conventional power plants.

Abbreviations and Symbols

Abbreviations

d.c.	=	diameter class	
ECU	=	European currency unit	[ECU]
FhG-ISE	=	Fraunhofer Institute for Solar Energy Systems	
grp	=	glassfibre reinforced plastics	
HAR	=	horizontal axis rotor	
MAWS	=	mean annual wind speed	
na	=	data not available	
O+M	=	operation and maintenance	
WEC	=	wind energy converter	
VAR	=	vertical axis rotor	

Countries

B	=	Belgium
DK	=	Denmark
E	=	Spain
F	=	France
FRG	=	Federal Republic of Germany
GR	=	Greece
I	=	Italy
IRL	=	Ireland
L	=	Luxembourg
NL	=	The Netherlands
P	=	Portugal
UK	=	United Kingdom

External Countries

AU	=	Austria
CH	=	Suisse
J	=	Japan
S	=	Sweden
USA	=	United States of America

Technical parameters of a WEC

a	=	year		
A	=	effective rotor swept area	$= D^2 * \pi/4 * 0.98$	[m²]
D	=	rotor diameter		[m]
p	=	specific rated power	$= P / A$	[W/m²]
P	=	rated power		[kW]
H	=	hub height		[m]
L	=	lifetime expectancy		[years]
v_{cut-in}	=	cut-in wind speed		[m / s]
v_{rated}	=	rated wind speed		[m / s]
$v_{cut-out}$	=	cut-out wind speed		[m / s]
v_{surv}	=	survival wind speed		[m / s]

Abbreviations and Symbols (contd.)

Physical quantities in energy generation

C	=	scale factor in Weibull distribution	
K	=	shape factor in Weibull distribution	
c_p	=	power coefficient	
e	=	specific energy generation	[kWh / m^2 T]
e_0	=	specific energy gain potential	[kWh/m^2 a]
e_{0H}	=	specific energy gain potential at height H above ground	[kWh/m^2 a]
E	=	energy generation	[kWh / a]
f_R	=	Rayleigh wind speed propability	[%]
f_W	=	Weibull wind speed propability	[%]
F	=	capacity- or Load-factor	[%]
k	=	relatio of delay v_3/v_0	
p_0	=	power density of wind flow	[W/m^2]
v_0	=	initial wind velocity upstream of rotor	[m/sec]
v_∞	=	wind velocity in rotor plane	[m/sec]
v_3	=	reduced wind velocity downstream of rotor	[m/sec]
T	=	time	[years]
η_{cutin}	=	energy losses caused by v_{cutin}	[%]
η_{rated}	=	energy losses caused by v_{rated}	[%]
η_{cutout}	=	energy losses caused by v_{cutout}	[%]

Derived quantities in economics

C_E	=	costs for electrical grid connection	[ECU]
C_{EW}	=	costs ex works	[ECU]
C_F	=	costs for foundation	[ECU]
C_M	=	costs for miscellaneous	[ECU]
C_T	=	costs for transport	[ECU]
C_{ER}	=	costs for erection	[ECU]
e_{inv}	=	annual energy output per unit of investment	[kWh/ECU]
i	=	annual interest rate on capital	[%]
g	=	generation costs	[ECU/kWh]
K	=	annuity factor	[%]

Derived quantities in energy pay-back

R_A	=	annual harvest factor	
R	=	harvest factor	
T_A	=	energy pay-back time	[years]

1 INTRODUCTION

The following report presents an analysis of the progress of wind energy technology in Europe from January 1986 to December 1989. The data used herein origin from the EUROWIN database, a project of the CEC-DG XII Wind R&D-programme, which contains data from wind energy converters (WECs) which are in operation throughout Europe. Data from EUROWIN are presented in various figures and tables in order to show the most important technological development and trends of wind energy technology in Europe between 1986 and 1989.

The aim of the present work was to produce a reference book for practical use. It is intended to benefit wind turbine manufacturers, public authorities, scientific research institutions and private investors interested in wind energy technology. The discussions herein are based on the fundamental concepts of fluid dynamics as applied to wind energy use. A statistical and graphical analysis of wind energy technology based on operational data, design parameters and economics is presented. The conclusions in this paper are based on the real data received from the operators and manufacturers of WECs throughout Europe over this four year period.

A collection of accurate wind energy data requires an accurate data source. A database and all results inferred from it are only as good and reliable as the data upon which it is based. All of the original input data were checked for plausibility before adding them to the database. Data outside certain limits were individually checked and if necessary corrected. However, the upper and lower limits for operational data set for each plant have been subject to our intuitive judgement. Although many incidental and systematic data faults can be found by this procedure, nevertheless an absolute guarantee of the correctness of every single data point cannot be given.

Technical terms, symbols and abbreviations that are used frequently throughout the text are defined on page iii. In section 2 a brief outline of the EUROWIN database organization is given. The main body of the report appears in sections 3 to 9. Here the results and evaluation of four years' data collected in the database are presented. Section 3 gives a brief survey over the development and status of wind energy technology in Europe and over the evaluation criteria, the results presented here are based upon. Section 4 treats the fundamental physics of energy conversion by WECs which is useful to compare the theoretical gains with the practical results in the sections following. Section 5 is a systematic representation of the most important design parameters and their development over 4 years. The master tables and graphics shown in this section give back the majority of technical information, which has been needed to build up the database.

In section 6 the most important results of 'Performance of European Wind Turbines' regarding energy generation, capacity factors and operational time are presented. Statistical failure analysis of various WEC-components, economics and last not least energy pay-back of wind energy use are treated in section 7, 8 and 9. The book is finished by a conclusion over the plenty of information given in the chapters before and a reference list of literature used and cited in the pages before in the sections 10 and 11.

In the appendix a more detailed description of the EUROWIN database is given followed by the derivations of mathematical algorithms as they are used regarding fundamental fluid mechanics, economics and energy pay-back. The annex also contains a catalog of WECs and manufacturers throughout Europe.

The use of wind energy, as one of the renewable energy sources, is not only a question on the power output, wind energy gain potential, generation costs and other technical and economic factors of wind energy technology.It also has a direct impact on the problem of our present energy hunger which is responsible for

- the rapid exploitation of limited fossil energy resources
- contamination and warming of the atmosphere by fossil power plants
- public concern about nuclear energy generation

This report treats only the above mentioned technical and economic aspects of modern WECs which are optimized for high energy generation and minimum costs to obtain generation costs which make wind energy use able to compete with fossil and nuclear power plants.

Wind energy is a viable alternative energy source and is a unique combination of tradition and modern technology. There is a vast potential for wind energy generation in the world and the technical development to use this potential is still in its initial stages. The conclusions in this report should serve as a framework on which to build further wind energy generation plants that are economically and technically efficient. We hope that this book will support the efforts of those who are striving towards economical as well as ecological energy generation.

2   Presentation of the European Wind Turbine Database EUROWIN

The EUROWIN database project was set up by the Commission of the European Communities DG XII in August 1985 under a project called "Databank on Existing Wind Turbines and Wind Climates in the Community". The objective of the work was to develop a database which provides information and statistics of current wind turbine technology and wind turbine sites in Europe. After five years the EUROWIN wind turbine database has become increasingly well known to institutions, manufacturers, publications and organisations engaged in wind energy research.

To date, manufacturers and operators in eight European countries have contributed information and data to EUROWIN. The database contains operational data from approximately 3200 wind turbines from various sites and of different sizes. This corresponds to a total installed wind energy capacity of nearly 320 MW in the European Community. From this collection of data, relevant information on the technology and economics of operating wind turbines in the European Community can be summarized.
The most important information gathered from the database concerns the technical data, installation costs, energy production and failures experienced by the operating WECs. The data acquisition itself is carried out in each country at the site by a national sub-contractor, who completes a standard questionnaire and a monthly postcard. The information gathered is transferred back to the database continuously. In the Netherlands and Denmark, where national databanks with operational WEC data already exist, the information for EUROWIN is obtained directly from these databanks.

The following organisations, institutions and consultants have supported this project :

Country	Name & address	Data delivery
Belgium	Mr. T.Vyslouzil HMZ Industriezone V Rellestraat 3 3800 St.Truiden	Reports
Denmark	Mr. T.Moller, Mr.W.Canter Naturlig Energi Vrinners Hoved 8420 Knebel	National database
Germany	Mr.J.P.Molly DEWI Ebertstr. 96 2940 Wilhelmshaven	Questionnaires
	Mr.U.Stampa Borchshöher Feld 23 2820 Bremen 70	Questionnaires
	Mr. W.Eggersglüß Landwirtschaftskammer Schleswig Holstein Postfach 1112 2300 Kiel 1	Reports
	Mr. Helmut Häuser Ingenieurwerkstatt Grundstr. 17 2000 Hamburg 20	National database

Country	Name & address	Data delivery
Greece	Prof.D.P.Lalas C.R.E.S. 6, Frati Street Fousia 19400, Koropi	Questionnaires
Ireland	Mr.L.D.Staudt Hurley Staudt Associates 2 Mary Street Drogheda Co. Louth	Questionnaires
The Netherlands	Mr.H.Berghuizen Communicatie-en Adviesbureau Leeuwenstraat 9-11 Postbus 21421 3001 AK Rotterdam	National database
Spain	Prof.E.M.Alonso, Mr. Sanchez I.E.R., C.I.E.M.A.T. Avda. Complutense, 22 D.P. 28040 Madrid	Questionnaires
United Kingdom	Dr.A.Garrad Garrad Hassan and Partners 9-11 St Stephens Street Bristol BS1 1EE	Questionnaires
CEC Programme Manager	Dr.Giancarlo Caratti Commission of the European Communities Directorate-General XII 200 Rue de la Loi Brussels	

The development of the EUROWIN database can be divided into **three** phases.

The **first phase** of the project consisted of choosing the information to be contained in the database. To accomplish this task, questionnaires and a monthly postcard for collecting technical and operational data were written and then distributed by the national sub-contractors. For a sample of this questionnaire, see Appendix: questionnaires. In 1985 when EUROWIN was established, there were about 750 registered WECs. The majority of these were in Denmark and the Netherlands. Today in 1991, the number of registered WECs in the database has more than tripled.

In the **second phase** the structure of the EUROWIN database was determined and programmed on a host computer. Many changes were made as experience was gained in evaluating the data. For example, all questions about tariffs of the local electricity utilities, which initially were asked in the questionnaires, were omitted from the later phases. Furthermore, in the first two years, much information was gained on failure analysis. Certain components were recognized to fail more often and more detailed information was sought in these areas.

The **third phase** consists in continuous evaluations of the operational data. Conclusions, summaries, statistics and general trends, which have been drawn from this data, have been presented in published form in some international wind energy journals, see Schmid(1986); Klein (1986); Schmid, Klein, Godard (1988); Schmid, Klein (1989); Schmid, Klein (1990); Schmid, Klein, Hagedorn (1991) and at international wind energy conferences, see Schmid (1984); Schmid, Klein (1987); Schmid, Klein, Godard (1988); Schmid, Klein (1988), Schmid, Klein (1990).

3 General Survey of Wind Energy Technology in Europe

3.1 Status of the Database

This chapter presents an overview of the development of wind energy technology in Europe from 1986 to 1989 as gathered from the EUROWIN database. Various information on WECs in each member country is available from the database, and the total data registered increases monthly. Table 1 below gives the status of the database, an update can be plotted by a standardized mask at any time, see examples in Appendix A. The table summarizes the WECs and manufacturers who contributed to EUROWIN until now. In the first two columns of Table 1 the number of WEC manufacturers and the number of different WEC-types, registered between 1986 and 1989 for each country, are listed. WECs with a rated power below 200 W do not appear in this table.

table 1: Participants in the EUROWIN database for contributing EC countries, including some WEC manufacturers from external countries

country	Manufacturers number	WEC types number	country	plants and units in operation plants	units	capacity (MW)
community :						
B	2	16	B	1	23	4.6
DK	73	373	DK	1743	2268	250.77
E	9	17	E	9	19	0.508
F	6	10	FRG	366	505	68.545
FRG	59	96	IRL	14	14	0.414
GR	1	8	NL	399	416	53.229
I	9	12	UK	13	13	1.701
IRL	4	5				
L	0	0				
NL	33	131				
P	0	0				
UK	22	50				
external :						
AU	1	2				
S	2	3				
USA	2	3				
CH	1	4				
J	1	3				
total :	234	765	total :	2597	3258	389.768

update : 12.07.1991

The data from some external WEC manufacturers obtained from technical reports have been evaluated only in certain statistics relating to design parameters; they do not appear in any statistics concerning operational data.

Table 1 shows the status of the database with update 14.06.1990 which is not identical with the status of wind energy technology at that time. For example in Germany 59 WEC manufacturers and 96 WEC types were registered: this is because one manufacturer can produce several WEC types. Similary 87 WEC plants and 126 individual WECs were registered. One plant can consist out of several operating units (windparks), hence in general the number of WECs is higher than the number of operators.

Since 1988 manufacturers in the higher power classes have become more specialized so that a concentration in WEC manufacturers is observed. Some former WEC manufacturers have specialized in producing WEC components such as rotor blades, towers, electronics, etc. Table 1 does not take these effects into account. This table is purely a summation of the total number of manufacturers, prototypes and commercially produced WEC types for which technical information has been available during the years 1985-1989. The database contains information ranging from small WECs such as battery chargers with 200 W rating to large machines with 3000 kW rating.

In the third column, the number of WEC operators, the number of individual WECs in operation and the installed wind turbine capacity registered for each contributing country are listed.

Not all WEC operators deliver continous monthly data, therefore the reporting quote differs monthly and this will be treated in the next section. The name commonly used for a wind energy generation plant with more than one WEC is 'windpark'. Each windpark appearing in this statistics is registered as one operator or one plant.

Denmark The Netherlands and Germany are the largest manufacturers of WECs and have the highest installed capacity in the European Community. The erection and evaluation of the first conventional modern WECsin especially in Denmark and The Netherlands occured more than 10 years ago which may explain the relatively high registration quote.

In Germany the installed capacity is higher than indicated in Table 1 because there are five newly installed windparks that do not yet appear in this statistic. Negotiations to promote a continuous data flow from these plants are still in progress.

In Spain and Greece most of wind energy activities remained in R&D stage before 1989. The European Wind Atlas shows that both countries have a relatively high wind energy potential. A typical application in these two countries is to observe since beginning 1990 and an accompanying evaluation can be expected in the future.

In France, Italy, Luxembourg and Portugal no notable wind energy activities are to observe which may change at least for Portugal and Italy in future. In France this may be explained by the state energy policy, although there is a high wind energy potential especial in the provinces of Bretagne, Normandie and Rhòne.

To date all the evaluated operational data have originated from windturbines of the horizontal axis type. The technical data of vertical axis machines are stored in the database regarding their technical data.

3.2 Evaluation Criteria

The results of evaluations made between 1986 and 1990 which are presented in the following chapters are based on certain evaluation criteria. The diagrams containing operational results do not allow a re-identification of WECs in order to safeguard commercially sensitive information about operating WECs and to avoid any preferential or disadvantageous treatment of a manufacturer. Exceptions were made for well-known machines such as NIBE A and NIBE B, GROWIAN which appear by name in some paragraphs.

The data presented can be dividedinto two different groups :

1. individual data : data of individual WECs mostly presented in point diagrams which show the scatter and the statistical distribution of a certain quantity

2. average data : data of averages mostly presented as histograms, curves or stapled areas obtained from certain groups of WECs such as diameter classes, power classes and country, or as averages over a certain time period . The graphs show general tendencies and developments

Fig. 1 shows the principle of presentaion for individual and average data (y) as function of x with parameter P.

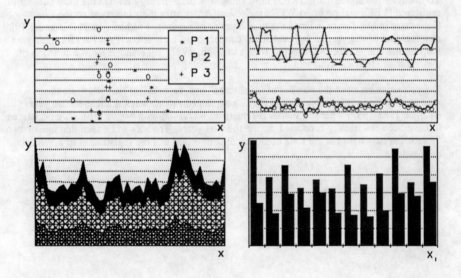

Fig. 1:   Principle of presentation for individual and average data

The individual data, which are the basis of the average data, further are classified as

- registered
- reported

WEC data. This distinction is important in analysing data from the database.

A WEC is considered to be 'registered' when its technical and site data such as rated power, rotor diameter, location, local wind conditions, etc. are made available from the operator. The survey of registered WECs only covers the fixed data of operating WECs, it does not include operational data.

A WEC is considered to be 'reported' when at least one monthly report in a given year, including any operational information, has been received by the database. Not all reported WECs include all the information needed for certain evaluations such as energy generation, failure analysis, economics, etc. because of incomplete monthly reports. The number of reported WECs for which checking of the procedure of data collection, inputting and plausibility of data etc. could be carried out, depends on the kind of evaluation and therefore differs for each diagram.

When analysing data that include mean values such as mean specific energy, capacity factors, etc. these averages are based on the reported WECs for a given year and they are weighted with the total number of WECs in the case of windparks.

To illustrate the above distinction, Fig. 2 shows the discrepancy between the total number of registered WECs and reported WECs for several years after their electrical connection. Outer bars in this figure indicate the total number of registered turbines as of December 1989 as a function of the year of their electrical connection. The four enclosed bars for each year indicate the real number of reported WECs in 1986, 1987, 1988 and 1989 as a function of their year of electrical connection.

Discrepancies between the number of 'registered' and 'reported' WECs illustrate the reporting quote as it is in reality and may be explained in two ways. First there are years (such as 1985-1989) where in successive years an increasing number of WECs are registered to have been electrically connected during the past years. Thus it can be seen that one or two years often pass between the time of electrical connection and the registration of a WEC in the database.

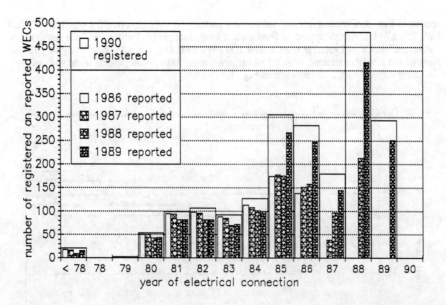

Fig. 2:   registered and reported turbines as a function of their year of electrical connection

For example the outer bar for the year 1987 indicates a total number of about 170 'registered' WECs to be electrically connected. In that year, however, only 40 WECs were 'reported' and evaluated to have been electrically connected. In the next year 1988 , the number of WECs reported to the database that claim to have been electrically connected in 1987, rose to approximately 100. This shows an increase of 60 reported WECs from 1987 to 1988 which indicate 1987 as the year of their electrical connection.

A maximum for registered and reported WECs is observed for 1985 which means that 1985 must have been a "boom year" for newly connected WECs in Europe.

Secondly there are years when the number of WECs that were reported to be electrically connected in a certain year has decreased in the following years (eg. 1977-1985). An explanation for this trend may be that after 3-4 years of continuously reporting data, some operators lose interest in delivering monthly data and cease to report data to EUROWIN.

Overall, it is important to note that there are more registered WECs per year than are reporting data to EUROWIN. In general, there has been a continuous increase in the number of WECs that are electrically connected since the beginning of the 1980's; however, the number of WEC operators who report data to the EUROWIN database each month fluctuates.

In Fig.3 one can see the rated power of 'registered' WECs as a function of their individual date of electrical connection. Note that windparks appear here as one point which represents the rated power of one machine. A few WECs which were connected before 1978 and those where the date of connection could not be made available do not appear in this figure. Overall there are about 2000 'registered' WECs stored in the database of which nearly 1700 appear in Fig. 3. Certain typical rated powers such as 30 kW, 55 kW, 75 kW, 100 kW, 200 kW, 250 kW and others appear as lines over a long time range. Before 1987 the majority of connected WECs had rated powers below 100 kW; however, since the beginning of 1988 most of the newly connecetd WECs had ratings above 100 kW. A general trend toward installing bigger WECs can be seen in Fig. 3.

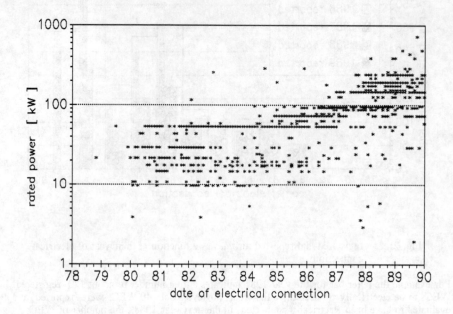

Fig. 3 :   Rated power of registered wind power plants versus individual date of electrical connection

Fig. 4 shows the development of the EUROWIN database regarding the number of reported WECs and the corresponding installed capacity for each month between 1986 and 1989. The gap between number of WECs and installed capacity becomes smaller with time which means that the newly connected WECs had in general a higher rated power, a fact that will be discussed later, see Fig. 34 in section 5.5. The fall in September and October 1989 means that there was an unusually low reporting quote in those months.

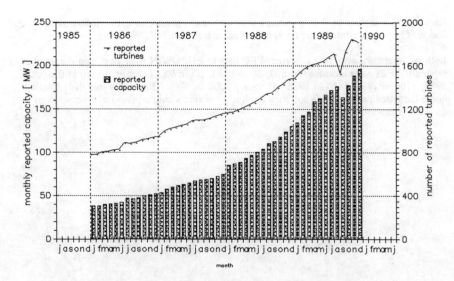

Fig. 4:   Reported WECs and corresponding installed capacity within the
            EUROWIN database

Fig. 5 distinguishes between the number of reported operators which is equal to the reported
plants, the reported WECs (single units) and the newly connected WECs in the years 1986,
1987, 1988 and 1989.

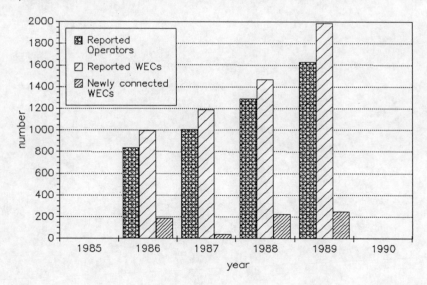

fig. 5:   Total number of reported operators, WECs and newly connected WECs
            for 1986-1989

Each year the number of WECs exceeds the number of operators due to windfarms in which one WEC operator controls the operation of more than one WEC.

Both the operators who have reported WEC data to EUROWIN each year, and the number of reported WECs have increased from 1986 to 1989. 1000 WECs were reported in the year 1986, 1200 in 1987, 1450 in 1988 and nearly 2000 in 1989. This means that the reporting quote increased progressively to twice the inital number within a period of four years of database operation.

4 Principles of Energy Generation by Modern WECs

Freely flowing wind can be used for energy generation by modern WECs which convert the kinetic energy into electrical energy. The rotor blades collect the kinetic energy over the rotor swept area and transform it into mechanical energy which is converted into electrical energy by a generator. The fundamental fluid mechanics relating to wind energy conversion are briefly reviewed. The most important links in wind energy use will be presented in the following figures.

Modern WECs are designed to generate electrical energy with high technical availability and low capital and operational costs. The technical development of wind energy technology involves an optimization of each link between the energy input by the rotor blades and the energy output by the generator. Energy production by WECs must be economically viable to compete with conventional energy conversion systems.

The chain between the intial energy source, the wind, and the effective generation costs of wind energy are schematically illustrated in Fig. 6, which shows the most important links.

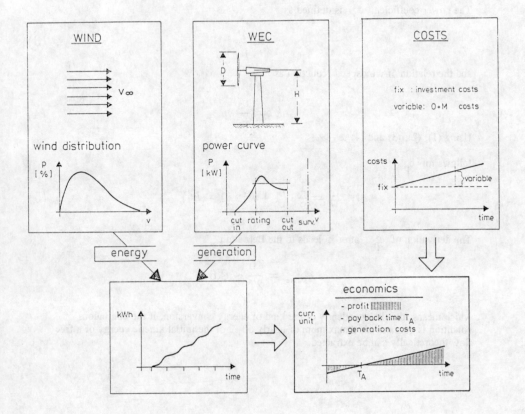

Fig. 6: The most important links between the initial wind energy source and generation costs for operational WECs

4.1   Energy Gain Potential

The power density of wind, the first link in this chain, and its theoretical maximum physical conversion according to the Betz limit, see equ. (1) and (6), is shown in the Fig. 7.

A brief abstract of the most important physical connections in fluid dynamics shall be given in this section.
The power density p_0 of a freely streaming fluid follows the equation

$$p_0 = \frac{1}{2} \cdot \rho \cdot v_0^3 \tag{1}$$

A rotor, placed in a free flow, reduces the intial windspeed v_0 to v_3 far behind the rotor plane. It follows from the Frouds actuator disc theorem for the windspeed in the rotor plane

$$v_\infty = \frac{1}{2} \cdot (v_0 + v_3) \tag{2}$$

The power coefficient c_p is defined as

$$c_p := \frac{p}{p_0} \tag{3}$$

and the relation of windspeed reduction as

$$k := \frac{v_3}{v_0} \tag{4}$$

Using (1), (2), (3) and (4)

follows for c_p

$$c_p = \frac{1}{2} \cdot (1 + k) \cdot (1 - k^2) \tag{5}$$

The derivation of c_p after k leads to the Betz limit

$$c_{pmax} = \frac{16}{27} \sim 0.6 \tag{6}$$

which means, that independently of the kind of energy conversion, if the continuous condition is guaranteed, a maximum of nearly 60 % of the initial kinetic energy of a free flow theoretically can be extracted.

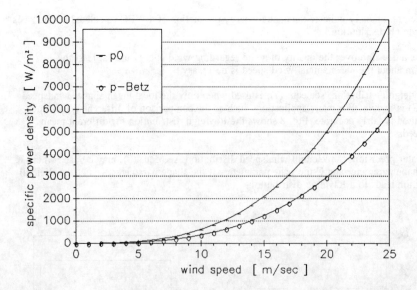

Fig. 7:   Power density of air and its maximum conversion after Betz limit as a function of windspeed

The power density curve increases with the cube of the windspeed. Depending on the wind conditions for which the WEC is optimized, the broken line indicates the maximum specific rated power [W/m²] which cannot be exceeded according to the Betz limit. In Figures 16, 17, 18 and 19 the specific power curves of several WECs are plotted as a function of windspeed at hub height.

The wind does not flow continuously at a constand speed, but has a certain distribution in time and space. The wind speed spectrum approximately can be described by two models, which indicate the propability of certain wind speeds within a time period T

a) Rayleigh distribution

$$f_R(v) = \frac{\pi}{2} \cdot \left(\frac{v}{\bar{v}^2}\right) \cdot e^{-\frac{\pi}{4}\left(\frac{v}{\bar{v}}\right)^2} \tag{7}$$

where   v   is the mean windspeed within a time period T

b) Weibull distribution

$$f_W(v) = \frac{K}{C} \cdot \left(\frac{v}{C}\right)^K \cdot e^{-\left(\frac{v}{C}\right)^K} \tag{8}$$

where K is the shape parameter, and C the scale parameter

The K and C parameter depend from the location and the surface roughness of the terrain.

Manufacturers mostly indicate the annual energy procuction of a WEC calculated on the basis of a Rayleigh distribution.

To make a decision over the quality of a site regarding wind energy use, at least the indication about the mean annual wind speed is necessary.

The Rayleigh distribution, see equ. (7), is used when only the local mean annual windspeed (MAWS) is known. For many applications such as the comparison of different WEC types this method is fairly accurate. Fig. 8 shows the Rayleigh distribution for different mean windspeeds.

For greater precision, the Weibull windspeed disribution, see equ. (8), may be used where K is the shape parameter and C the scale parameter. If the shape parameter K is 2, the Weibull distribution leads to a Rayleigh distribution.

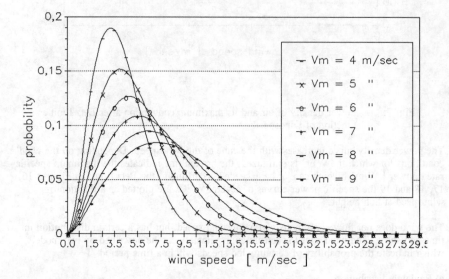

Fig. 8:   Rayleigh distribution for different mean annual windspeeds v_m

Note that the windspeeds on the horizontal axis in Fig.8 represent windspeed classes, e.g. 17.5 [m/sec] is the wind class from 17-18 [m/sec].

The integration of power over time leads to the energy in a certain period, see equ. (12). Based on a Rayleigh distribution the specific energy gain potential in [kWh/m² a] for a period of one year after equ. (14) leads to the graph shown in Fig.9

The energy is obtained by integration of power over time, using equ. (1) the specific energy per square metre is

$$e(t) = \int_0^T p \, dt \qquad (9)$$

Using a certain wind speed distribution $f(v)$

dt in (9) is

$$dt(v) = f(v) \cdot dv \cdot T \tag{10}$$

and (10) in (9) leads to

$$e(v) = T \cdot \int_0^v \cdot p \cdot f(v) \cdot dv \tag{11}$$

Using the Rayleigh distribution $f_R(v)$, (11) is

$$e_R = \rho \cdot \frac{\pi}{4} \cdot \frac{T}{\bar{v}^2} \cdot \int_0^v v^4 \cdot e^{-\frac{\pi}{4}\left(\frac{v}{\bar{v}}\right)^2} dv \tag{12}$$

The same for the Weibull distribution leads to

$$e_W = \frac{\rho}{2} \cdot \frac{K}{c^K} \cdot T \cdot \int_0^v v^{K+2} \cdot e^{-\left(\frac{v}{c}\right)^K} dv \tag{13}$$

the solution of (12) , after integration can be given as

$$e_R = \frac{3 \cdot \rho \cdot T}{\pi} \cdot \bar{v}^3 \tag{14}$$

derivation of (14) see Klein (1986)

The increase of velocity with height above ground can be indicated by the 1/7-law

$$v_H = v_{10} \cdot \left(\frac{H}{10}\right)^{\frac{1}{7}} \tag{15}$$

where v_H is the velocity at height H above ground

and v_{10} is the velocity at 10m above ground

Substituting v in equ. (12) with v_H from (15) this leads to the energy gain potential in height H, to see in Fig. 10.

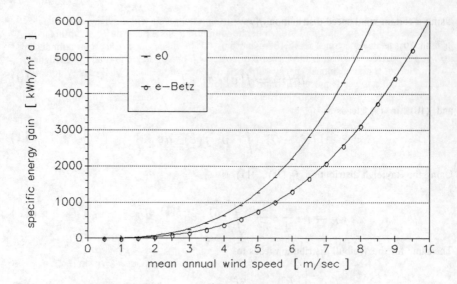

Fig. 9: Specific energy gain potential as a function of mean annual windspeed

As for power, the energy E_Q has a maximum for conversion after the Betz limit.
The usual height for measuring the mean site annual windspeed is 10m above ground.
Because of the surface roughness a boundary layer is formed in the vertical axis. On the basis
of a Rayleigh distribution, the influence of the height above ground can be plotted as shown
in Fig. 10.

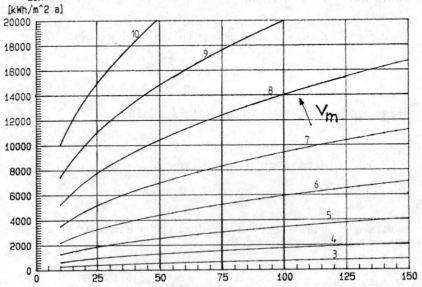

Fig. 10: Specific annual energy gain potential as a function of height H above
ground for different mean annual windspeeds v_m at 10m above ground,
calculated on the basis of a Rayleigh distribution and 1/7 power law, taken
from Klein (1986)

As expected, the influence of the height above ground is greater with higher mean annual windspeeds. Most new machines today operate with hub heights between 30 and 50m above ground, large scale machines reach 100m and more at the blade tip. For the optimum design of a WEC one has to consider the increase of costs with height above ground. With higher towers there is a higher annual energy output. It depends on the local wind energy potential whether a higher tower is more economic than a lower one.

The relationship of hub height to the specific energy output of operational windturbines shall be treated later in section 6.16.

5 Design Parameters

5.1. Systematic Representation of WECs

Master Table 2-6 which follow present a survey of the technical data of more than 600 different WECs. Tese tables contain all the information which has been made available from technical reports since 1985 whether or not a WEC is still being produced by the manufacturer. The following key-table below describes the codes used in the Table 2-6.

- binary codes   (*) :   various combinations possible
- numeric codes   (#) :   only one combination possible

Wind energy converters are classified in the following rated power ranges :

class I	:	0.1 -	10 kW	rated power	see Table 2
class II	:	10.1 -	75 kW	"	see Table 3
class III	:	75.1 -	300 kW	"	see Table 4
class IV	:	300.1 -	900 kW	"	see Table 5
class V	:	>	900 kW	"	see Table 6

The WECs are listed by country. The order of countries is as shown in the list of abbreviations and symbols, see page iv.

fields :

field	content
CNT	country
MANUF	manufacturer
RATPOW	rated power
DIAM	diameter
SPECPOW	specific rated power
PROFILE	section profile
WEIHEAD	weight of tower head
VCUTIN	cut-in wind speed
VRATED	rated wind speed
VCUTOUT	cut-out wind speed
VSURV	survival wind speed
TOWHEI	tower height

* binary codes :

field	content	1	2	4	8	16	32	64	128	256	512
POWCONT	power control	stall	pitch	part. pitch	flex. blade	el-hydr.	mechanic	var. geom.	yawing	side vane	ecliptic
OVERPROT	overspeed protection	stall	pitch	mech. br.	wing tips	flaps	parachute	eddy-curr.	ballast res.	side vane	ecliptic
GENTYP	generator type	synchron	asynchr.	DC	1-phase	2-phase	AC-DC-AC	others	perm. magn.		
TRANS	transmission type	spur gear	planet gear	other gear	belt	direct	hydraulic	chain			
BRKTYP	brake type	disc br.	drum br.	strap br.	no brake						
TOWTYP	tower type	tubular	cantil.	guyed	lattice	flexibel	no tower	tilting			
TOWMAT	tower material	steel	concr.	wood	others						

* numeric codes :

field	content	1	2	3	4
ROTTYP	rotor type	horizontal	vertical		
ORIENT	orientation	upwind	downwind		
SPEED	rotor speed	fix	variable	1-speed	2-speed
BRKPOS	brake position	low drivetrain side	fast drive train side		

Table 2 : WECs in the rated power range 0.1 - 10 kW

CNT	MANUF.	TYPNAME	RATPOW	DIAM	SPECPOW	ROTTYP	NUMBLAD	ORIENT	POWCONT	SPEED	PROFILE	WEIHEAD	VCUTIN	VRATED	VCUTOUT	VSURV	GENTYP	TRANS	BRKTYP	BRKPOS	TOWTYP	TOWMAT	TOWHEI
			kW	m	W/m²	#	#	#	#	*	#	kg	m/sec	m/sec	m/sec	m/sec	*	*	*	#	*	*	m
B	TURBO	V6	3,00	6,0	108,3	2	2	3		34			3,0	10,0									
B	TURBO	V86	8,00	8,0	162,6	2	2	3		34			3,0	10,0									
B	HMZ	10kW8	10,00	8,0	203,3	1	1	1		1													
DK	OESTAS	Oestas 2 kW	2,00	5,0	104,7	1	1			34	4												
DK	WINCON	W 3	3,00	3,7	280,4	1	12	2	2	1	4						1	4			5	1	9,0
DK	BALTIC	3 KW	3,00	3,4	337,1	1	3	2	2	34	1						2						
DK	REYMO	REY 3,5 kW	3,50	5,0	183,2	1	2	1		2	1	195	2,0	8,0			16	1		2	5	1	12,0
DK	VIMOEL	VIMOE 4/5.4	4,00	5,4	174,7	1	3																
DK	DWT	WINDANE 5/4	4,00	5,0	204,1	1	2	1		18			4,5			80,0						1	12,0
DK	VIBLIS	VIBL 7/6	7,00	6,0	247,3	1	3		1	1													
DK	THYMOE	THY 7.5/7.5	7,50	7,5	169,7	1	3		1	1													
DK	VIMOEL	T 103	10,00	7,0	263,2	1	3		1	1													
DK	VIMOEL	VIMOE 10/7.5	10,00	7,5	226,2	1	3		1	1							1				9	1	15,0
DK	SJ WIN	SJ 10/7	10,00	7,0	259,7	1	3		1	1													
DK	GSS	GSS 10/6.1	10,00	6,1	342,5	1	3		1	1													
DK	NOTANK	NTK 10/8.3	10,00	8,3	182,8	1	3		1	1													
DK	T 103	T 103 10/8	10,00	8,0	199,2	1	3		1	1													
DK	WINFOS	10/7.5	10,00	7,5	222,2	1	3		1	1													
E	PROENE	60.7	0,10	1,3	76,9	1	2	1		34			2,5	7,0			64						
E	BORNAY	G200	0,20	2,0	64,5	1	2	1		34			3,0	10,0			512	16			9		
E	PROENE	150.7	0,20	2,0	64,5	1	2	1		4			3,0	7,0			64	16			1		
E	HAZSOL	WPN-ZD 0.2	0,30	2,0	96,8	1	2	1		34			3,5	8,0			8	16				1	
E	GENZ	500 W	0,50	2,6	96,2	1	3	2		34			2,0	7,0			512	16					
E	HAZSOL	WPN-ZD 0.8	0,80	2,0	258,1	1	2	1		4			2,5	10,0			8	16			1	1	
E	BORNAY	G1000	1,00	5,0	52,1	1	3	1		34			3,5	12,0			512	16		2		1	
E	PROENE	1000.9	1,00	3,2	126,6	1	2	1		34			2,0	9,0			64	16			9	1	
E	GENZ	1800 W	1,80	3,6	180,0	1	4	2		34			2,0	7,0			512	16					
E	GENZ	2000 W	2,00	3,6	200,0	1	4	2		34			2,0	7,0			512	16					
E	BORNAY	G5000	5,00	6,0	182,5	1	3	1		34		250	2,0	5,5	12,0	60,0	1	1			1	1	9,0

Table 2 : WECs in the rated power range 0.1 - 10 kW (contd.)

CNT	MANUF.	TYPNAME	RATPOW kW	DIAM m	SPECPOW W/m³	ROTTYP #	NUMBLAD #	ORIENT #	POWCONT *	SPEED #	PROFILE	WEIHEAD kg	VCUTIN m/sec	VRATED m/sec	VCUTOUT m/sec	VSURV m/sec	GENTYP *	TRANS *	BRKTYP *	BRKPOS #	TOMTYP *	TOMMAT *	TOWHEI m
F	AERMAT	60-7	0,10	1,3	76,9	1	2	2	34				3,5	7,0			64				1	1	
F	AERMAT	150-7	0,20	2,0	64,5	1	2	2	34				3,5	7,0			64				1	1	
F	AERMAT	400-7	0,40	3,2	50,6	1	2	1	34				3,5	7,0			64	4			1	1	
F	AERMAT	UM70/2.5	2,50	7,0	66,3	1	2	1	34				4,2	9,0			4	4			1	1	
F	AERMAT	UM70/5	5,00	7,0	132,6	1	2	1	34								4	4			1	1	
F	AERMAT	UM70/10	10,00	7,0	265,3	1	2	1	34				5,3	11,0			4	4			1	1	
FRG	WIKZ	HD 06512	0,10	1,4	66,7	1	3	1	1				4,0	8,0			512	16			5	1	
FRG	BRUEM	BW08	0,10	0,8	200,0	1	3	1	1				4,0	8,0			128	16			5	1	
FRG	HARB	WG 910	0,10	0,9	166,7	1	6	1	1	4		14					128	16			5	1	
FRG	HARB	C. 100-12	0,10	1,6	47,6	1	4	1	1	4		32						4			5	1	
FRG	LUBING	M 015-4	0,10	1,5	55,6	1	2	2	1			35	3,0	8,0					8		16	1	6,0
FRG	SOVENT	W 100	0,10	1,1	90,9	2	3	3		4		40												
FRG	LUBING	ML 015-6-3	0,20	1,5	111,1	1	6	2	1			59	3,0	7,0			128	4	8		5	1	3,0
FRG	WIKZ	HD 250 W	0,25	1,8	100,0	1	3	1	1				2,5	9,5									
FRG	WIKZ	HD 312, HD 324	0,30	2,4	78,9	2	2	2	18	4		105	4,0	14,0	99,0						16		
FRG	SOVENT	W 300	0,30	1,8	85,7	2	2	3		4					99,0		2	1	1		16		
FRG	KAEHL	Kano-Rotor 600S	0,60	2,5	125,0	1	3	2	34	1			3,5	12,0	99,0		2	1	1		10	1	6,0
FRG	SUEWIN	B 25	0,70	2,5	142,9	1	2	2	34								4				5	1	8,0
FRG	SUEWIN	B 30	0,90	3,0	126,8	1	2	2	34								4				5	1	8,0
FRG	ENRTCH	SWEC2	1,00	2,0	322,6	1	2	2	1	1			3,0	13,2			4	16			5	1	
FRG	SCHUB	ES 03-01	1,00	3,0	140,8	1	2	2	1	4		205	3,0	10,0	20,0		2				5	1	6,0
FRG	HUELL	1 KW Anlage	1,00	2,6	192,3	1	2	2	1				4,0	13,0			8	4			3	1	
FRG	WIKZ	Elektromat 1 kW	1,00	3,0	144,9	1	3	1	513	4	NACA 4412		2,2	9,0	10,0	60,0	144	16	4		5	1	
FRG	SUEWIN	E 305	1,50	3,4	164,8	1	5	2	1				4,0	9,0	30,0	50,0	1	1			5	1	12,0
FRG	SUEWIN	E 305	1,80	3,4	202,2	1	5	2	1	4		200	3,0	9,0	30,0	67,0	5	24,0		1	5	1	18,0
FRG	SCHUB	WE 03-03	3,00	4,0	243,9	1	2	1	1	1							2	1	8		5	1	
FRG	SCHUB	WE 03-03	3,00	4,0	238,1	1	2	1	1			240	4,0	10,0	20,0		2	1			5	1	
FRG	VENTIS	04-03	3,00	4,0	245,9	1	2	1		4	NACA 44xx		3,5	12,0		60,0	2				5	1	12,0
FRG	BORN	S/5.5	5,50	5,0	282,1	1	3	1	1														

Table 2 : WECs in the rated power range 0.1 - 10 kW (contd.)

CNT	MANUF.	TYPNAME	RATPOW kW	DIAM m	SPECPOW W/m²	ROTTYP #	NUMBLAD #	ORIENT #	POWCONT *	SPEED #	PROFILE	WEIHEAD kg	VCUTIN m/sec	VRATED m/sec	VCUTOUT m/sec	VSURV m/sec	GENTYP *	TRANS *	BRKTYP *	BRKPOS #	TOWTYP *	TOMAT *	TOWHEI m
FRG	SUEWIN	THG 5	6,00	3,4	666,7	1	5	2	1	1		200	4,0	14,0	30,0		1	1		4	5	1	18,0
FRG	SUEWIN	E 710	8,60	7,0	228,1	1	3	2	1	1			3,0		30,0	67,0	17	1			5	1	30,0
FRG	SCHUB	WE 08-10	10,00	8,0	202,8	1	3	1	1	1				12,0				16				1	12,0
FRG	AEE	PG 10	10,00	6,3	331,1	1	3	2	34				3,0	10,5	25,0	50,0	128	16			3	1	18,0
GR	WISO	BW 1KA	1,70	3,0	242,9	1	3	1	90	4	NACA 4412	145	3,5	9,0	25,0		8	1	1		8	1	6,0
GR	WISO	BW 2KB	2,50	5,0	131,6	1	3	1	90	4	NACA 4412-24	185	3,5	9,0	25,0		8	1	1		8	1	9,0
GR	WISO	BW 3KB	3,50	5,0	184,2	1	3	1	90	4	NACA 4412-24	190	3,5	9,0	25,0		8	1	1		8	1	9,0
GR	WISO	BW 5KC	5,00	7,3	122,0	1	3	1	1	1	NACA 4415-28	600	3,0	9,0	20,0	50,0	1	1	1		8	1	12,0
GR	WISO	BW 10KCC	10,00	7,3	243,9	1	3	1	1	1	NACA 4415-28	680	3,0	11,0	20,0	50,0	1	1	1		8	1	12,0
GR	WISO	BW 10KCG	10,00	7,3	243,9	1	3	1	1	1	NACA 4415-28	720	3,0	11,0	20,0	50,0	1	1	1		8	1	12,0
I	CAST	3.4.5	1,00	4,0	81,3	1	3	1	34				2,0	9,0	9,0		512	16			8	1	
I	CEMEL	48211	3,00	1,8	1200,0	1	36	1	34				1,3				512	1					
I	RIVA	MP 5	3,60	5,3	166,7	1	1	2	18				5,0	12,0	18,0		1	1			1	1	9,0
I	RIVA	M 7	5,20	6,5	160,0	1	1	2	2	4	NACA 44xx		3,9	10,0	20,0	50,0	4	1			1	1	
I	ISEA	6/10	10,00	6,0	361,0	1	2	1	34				4,0	11,8								5		
IRL	JACO	Jacobs 1.8KW	1,80	3,7	171,4	2	3	4	1	1							2	8					
IRL	COMHAR	CCFI	3,00	3,0	434,8	2	3		1				5,0	14,0	14,0	50,0	900					1	5,0
IRL	PRIMA	1kW	3,00	3,0	434,8				1				5,0	12,5									
NL	LMW	LMW 150	0,15	1,5	88,2	1	3	2	128		Goettingen 417a	24	3,0	10,0		60,0	132	16			5	1	6,0
NL	LMW	LMW 250	0,25	1,7	113,6	1	3	2	257				3,0	10,0		60,0	8				1	1	
NL	LMW	LMW 600	0,60	2,2	157,9	1	2	1	257	4	Clark Y	52	3,0	12,0		60,0	132	16	8		5	1	
NL	LMW	LMW 1000	1,00	2,5	208,3	1	3	1	257	1	NACA 4412		3,2	13,0		60,0	64				1	1	
NL	LMW	LMW 1003	1,00	3,0	140,8	1	3	1	513			76	2,5	9,0		60,0	128	16			5	1	
NL	LMW	LMW 1000	1,00	2,5	208,3	1	3	1	257				3,2	12,0		60,0	8						
NL	LMW	LMW 1003	1,00	3,0	144,9	1	3	1	1		NACA 4412		2,5	10,0		60,0	8					1	

Table 2 : WECs in the rated power range 0.1 - 10 kW (contd.)

CNT	MANUF.	TYPNAME	RATPOW	DIAM	SPECPOW	ROTTYP	NUMBLAD	ORIENT	POWCONT	SPEED	PROFILE	WEIHEAD	VCUTIN	VRATED	VCUTOUT	VSURV	GENTYP	TRANS	BRKTYP	BRKPOS	TOMTYP	TOMMAT	TOWHEI
			kW	m	W/m²	#	#	#	*	#		kg	m/sec	m/sec	m/sec	m/sec	*	*	*	#	*	*	m
NL	LMW	LMW 2500	2,50	5,0	130,2	1	3	1	257		NACA 4412/18		2,0	12,0		60,0	144	16			5	1	18,0
NL	LMW	LMW 3600	3,60	5,0	183,7	1	3	1	257	1	NACA 4412/18		4,0	12,0		60,0	144	16			5	1	18,0
NL	LMW	LMW 3600(2 blad	3,60	5,0	188,5	1	2	2	1	1			4,0	12,0			128						
NL	LMW	LMW 10/7	10,00	7,0	265,3	1	3	1	257	4	BW 3	463	3,1	12,1		54,0	144	16			5	1	24,0
NL	H-ENSY	HE 600 L	3,00	6,0	108,3	1	2	1	4				4,5	12,0			2	4			1		
NL	H-ENSY	HE 1000 L	10,00	10,0	130,0	1	2	1	4				4,5	12,0			2	4			1		
NL	AIR	NN	5,50	6,5	170,8																		
NL	ELEKT	WV15	1,20	3,0	173,9	1		1	1														
NL	ELEKT	WV35	4,00	4,0	325,2	1			1														
NL	ELEKT	WV50	6,00	5,0	312,5	1			1														
NL	BERMOU	Windvang 65.8	8,00	6,5	246,2	1	3	1	34				4,0	11,0			2	2			8	1	
NL	B.C.A.	WTW 7	8,50	6,6	253,7	1			1												5	1	21,0
NL	FDO	5 WPX	9,00	5,0	459,2	1			1														
NL	POLENK	WPSBA10	10,00	8,5	176,1	1			1								2	2			1	1	17,0
UK	AMPAIR	100W	0,10	0,9	166,7	1	6	1	1				3,0				8				5	1	
UK	MARLEC	910 Series	0,10	0,9	166,7	1	6	1	1				1,8	9,8			8	16			1		
UK	MARLEC	1800 Series	0,30	1,8	120,0	1	3	1	1				1,8	9,8			8	16			5		
UK	MMEW	Vertax DD	0,10	1,2	90,9	2	3	1	1				2,0					4			1	1	
UK	AEROT	Model 2	0,20	2,0	64,5	1	2	1	1				2,6	15,0							1	1	
UK	AEROT	Model 3	0,40	2,0	129,0	1	4	1	1				2,1	15,0								1	
UK	MMEW	Vertax HD	0,40	3,0	58,0	2	3	1	1				2,0					4			5	1	
UK	TORNA	500 W	0,50	2,1	147,1	1		1	1				4,0	11,5			64	16			8		
UK	NATPOW	Vortex 200	0,20	2,0	64,5	1	2	1	1				4,0	9,0			2				8		
UK	NATPOW	Vortex 1000	1,00	2,0	322,6	1	2	1	1				3,0	12,0			2				8		
UK	PROVEN	Proven 1k	1,10	2,9	169,2	1	3	1	1														
UK	TWST	HAWK 4.2 mtr.	2,00	4,3	71,4	1	3	1	34	4	NACA 23018	200	4,0	8,0	20,0	60,0	8	9			5	1	7,0
UK	HAWKER	MP5	2,20	5,0	112,2	1	3	1	99	4	AEROFOIL		3,6	9,0	47,0	75,0	1	16			8	1	12,0
UK	ENSER	ES3	3,00	3,4	312,5	1	3	1	1	1	NASA LS1	170	5,0	13,0	30,0	56,0	1	1		1	1	1	7,0
UK	NATPOW	Vortex 6000	6,00	4,5	384,6	1	3	1	1				3,0	12,0			2	2			8		

Table 2 : WECs in the rated power range 0.1 - 10 kW (contd.)

CNT	MANUF.	TYPNAME	RATPOW kW	DIAM m	SPECPOW W/m²	ROTTYP #	NUMBLAD #	ORIENT #	POWCONT *	SPEED #	PROFILE #	WEIHEAD kg	VCUTIN m/sec	VRATED m/sec	VCUTOUT m/sec	VSURV m/sec	GENTYP *	TRANS *	BRKTYP *	BRKPOS #	TOWTYP *	TOWMAT *	TOWHEI m
UK	CRESS	CE.8000	8,00	7,5	184,8	1	3	2	34				3,5	10,0			4	1			8	1	
UK	CRESS	CE.8000	9,40	7,5	217,1	1	3	2	34				3,5	10,0			4	1			8	1	
UK	TRIM	Trimble 10 KW	10,00	6,0	361,0	1			1														
AU	VILLAS	Flair 8	5,00	8,0	99,4	1	1	2	1		FX 84-W serie										5	1	18,0
AU	VILLAS	Flair 8	7,50	8,0	149,1	1	1	2	65	1	FX 84-W serie		3,0	10,0	25,0	60,0	1	2			3	1	
USA	BERGEY	BWC 1000	1,20	2,8	200,0	1	3	1	9		4		4,0	13,0		54,0	128		8				15,0 / 37,0
USA	BERGEY	BWC 1500	1,50	3,2	189,9	1	3	1	1		4		3,1			54,0	128						12,0 / 30,0
USA	BERGEY	BWC EXCEL	10,00	7,0	265,3	1	3	1	9		4		3,4	12,1		54,0	128		8				15,0 / 37,0
CH	ELEKTR	Windgen.500W	0,50	2,5	102,0	1	2	1	2		4		2,5	8,5			144						
J	KIDEN	MSW-VAWT	8,00	2,5	1666,7	2	3		1		4		3,5	15,0	15,0								9,0

Table 3 : WECs in the rated power range 10.1 - 75 kW

CNT	MANUF.	TYPNAME	RATPOW kW	DIAM m	SPECPOW W/m²	ROTTYP #	NUMBLAD #	ORIENT #	POWCONT *	SPEED #	PROFILE	WEIHEAD kg	VCUTIN m/sec	VRATED m/sec	VCUTOUT m/sec	VSURV m/sec	GENTYP *	TRANS *	BRKTYP *	BRKPOS *	TOMTYP #	TOMAT *	TOWHEI m
B	HMZ	20KW12	20,00	12,0	180,5	1			1								2				1	1	17,0
DK	THYMOE	THY 11/8	11,00	8,0	223,6	1	3		1														18,0
DK	KURI	KURI 11/10.9	11,00	10,9	117,9	1	3		1														
DK	UKENDT	UKEN 11/10	11,00	10,0	140,1	1	3		1														
DK	WINMAT	WIMA 11/8	11,00	8,0	219,1	1	3		1														
DK	DWT	WINDANE 9/11	11,00	9,0	173,0	1	3		1														
DK	ALTERN	11/7.4	11,00	7,4	255,8	1	3		1														
DK	NORFOL	Folkecenter13.5	13,50	8,6	237,3	1	3		1	1													18,0
DK	KURI	KURI 15/10	15,00	10,0	191,1	1	3		1														
DK	POULS	POUL 15/10.5	15,00	10,5	173,4	1	3		1														
DK	REYMO	REY 15/8.5	15,00	8,5	264,6	1	3	1	1		NACA 63-xxx		3,6	15,0	30,0	66,0	132	2			8	1	24,0
DK	KRAMS	KRAM 15/8.3	15,00	8,3	279,9	1	3		1														
DK	KURREY	KURI 15/8	15,00	8,0	298,8	1	3		1														
DK	SOEG	Poulsen II	15,00	12,0	135,4	1	3		1														
DK	WINFOS	15/8.8	15,00	8,8	245,9	1	3		1														18,0
DK	KURI	KURI 18/11	18,00	11,0	189,5	1	3		1														
DK	REYMO	REY 18/8.5	18,00	8,5	317,5	1	3		1												10		
DK	OODER	ODD 18/11	18,00	11,0	189,5	1	3		1														
DK	DWT	DWT 9.6/18	18,00	9,6	249,0	1	3		1														
DK	BOST	Bost 18/10.9	18,00	10,9	152,5	1	3	1	1					14,0			2				10	1	18,7
DK	REYMO	REY 4/18.5	18,50	10,7	205,8	1	3	1	1	2	NACA 44xx	2075	3,0	15,0	30,0	66,0	18	2	1		10	1	24,0
																					3	1	24,0
DK	REYMO	REY 18.5 kW	18,50	8,5	332,7	1	3		1	1													18,6
DK	OODER	OM-18.5kW	18,50	10,9	202,4	1	3		1	1													14,3
																							18,3
DK	DWT	WINDANE 10/18	18,50	9,6	255,5	1	2	2	1	1	FX-W-151	500	4,5	15,0	70,0		4	1	1		5	1	18,0
DK	DWT	WINDANE 12/18	18,50	12,0	167,0	1	2	2	1	1													19,0
DK	BOST	Bosted Moellen	18,50	11,0	198,7		3																12,0
																							18,0
																							24,0

Table 3 : WECs in the rated power range 10.1 - 75 kW (contd.)

CNT	MANUF.	TYPNAME	RATPOW	DIAM	SPECPOW	ROTTYP	NUMBLAD	ORIENT	POWCONT	SPEED	PROFILE	WEIHEAD	VCUTIN	VRATED	VCUTOUT	VSURV	GENTYP	TRANS	BRKTYP	BRKPOS	TOWTYP	TOWMAT	TOWHEI
			kW	m	W/m²	#	#	#	*	#	#	kg	m/sec	m/sec	m/sec	m/sec	*	*	*	#	*	*	m
DK	DWT	WINDANE 12/20	20,00	12,0	177,0	1	2	1	1	1									1		5	1	18,0
DK	DANSME	SME 22/10(v.2)	22,00	10,0	286,1	1	3		1	1													
DK	NOTANK	NTK 22/11	22,00	11,0	231,6	1	3		1	1													
DK	KURI	15K	22,00	10,9	235,8	1	3		1	1													
DK	WINMAT	WIMA 22/10	22,00	10,0	280,3	1	3		1	1													
DK	HVK	HVK 22/12	22,00	12,0	194,7	1	3		1	1													
DK	RIJSAG	RIIS 22/10	22,00	10,0	280,3	1	3		1	1													
DK	KONGST	KONG 22/10	22,00	10,0	280,3	1	3		1	1													
DK	SONBJ	SONE 22/10	22,00	10,0	280,3	1	3		1	1													
DK	VEST/H	VEST 22/10	22,00	10,0	280,3	1	3		1	1													
DK	BONUS	BON 22/10	22,00	10,0	280,3	1	3		1	1							4	4	1	1		1	
DK	BONUS	BON 22/12	22,00	12,0	194,7	1	3		1	1							4	4	1	1		1	
DK	WINCON	M 22	22,00	9,8	282,1	1	3	1	1	1			4,0	12,0		50,0	131	1	1	1	1	1	18,0
DK	SOEG	SOEG 22/10.8	22,00	10,8	240,2	1	3		1	1													
DK	VENDEL	VEND 24/12	24,00	12,0	212,4	1	3		1	1													
DK	DWT	WINDANE 12/25	25,00	12,0	221,2	1	3		1	1													
DK	DANSME	SME 30/12	30,00	12,0	270,8	1	3		1	1													
DK	NOTANK	NTK 30/11	30,00	11,0	315,8	1	3		1	1													
DK	WINMAT	WIMA 30/12	30,00	12,0	265,5	1	3		1	1													
DK	WINMAT	WIMA 30/10	30,00	10,0	382,2	1	3		1	1													
DK	POULS	POUL 30/11	30,00	11,0	315,8	1	3		1	1													
DK	HVK	HVK 30/12	30,00	12,0	265,5	1	3		1	1													
DK	SONBJ	SM30	30,00	10,0	390,1	1	3		1	1													
DK	VEST/H	VEST 30/11	30,00	11,0	315,8	1	3		1	1													
DK	BONUS	BON 30/10	30,00	10,0	382,2	1	3		1	1													
DK	OR.WIN	OR 30/12	30,00	12,0	265,5	1	3		1	1							4	4	1		1	1	
DK	VESTAS	VEST 30/10	30,00	10,0	384,6	1	3		1	1													
DK	DWT	WINDANE 12/30	30,00	12,0	265,5	1	3		1	1													
DK	ALTERN	30/10.8	30,00	10,8	326,1	1	3		1	1													
DK	WINFOS	30/11.2	30,00	11,2	315,8	1	3		1	1													
DK	IRAS	VB-11	30,00	11,0	322,2	1	3		1	1			4,0	12,0				2			5		
DK	VESTAS	VESTAS 30/18	30,00	18,0	121,6	1	3		1	1													

Table 3 : WECs in the rated power range 10.1 - 75 kW (contd.)

CNT	MANUF.	TYPNAME	RATPOW kW	DIAM m	SPECPOW W/m²	ROTTYP	NUMBLAD #	ORIENT #	POWCONT #	SPEED *	PROFILE #	WEIHEAD kg	VCUTIN m/sec	VRATED m/sec	VCUTOUT m/sec	VSURV m/sec	GENTYP *	TRANS *	BRKTYP *	BRKPOS #	TOWTYP *	TOWMAT *	TOWHEI m
DK	RIJSAG		35,00	14,0	227,4	1	3			1									1				
DK	HVK	HVK 37/12	37,00	12,0	327,4	1	3			1									1				
DK	DWT	WINDANE 12/40	40,00	12,0	354,0	1	3			1									1				
DK	NOTANK	NTK 45/15	45,00	15,0	254,8	1	3			1									1				
DK	WINMAT	WIMA 45/14	45,00	14,0	292,4	1	3			1									1				
DK	SONBJ	SM45	45,00	12,0	406,1	1	3			1							4	4	1		1	1	
DK	BONUS	BON 45/15	45,00	15,0	254,8	1	3			1									1				
DK	ALTERN	45/12.5	45,00	12,5	365,9	1	3			1									1				
DK	DWT	WINDANE 12/50	50,00	12,0	442,5	1	3			1									1				
DK	DANSME	SME 55/16	55,00	16,0	279,3	1	3			1									1				
DK	NOTANK	NTK 55/14	55,00	14,0	357,4	1	3			1									1				
DK	NOTANK	NTK 55/16	55,00	16,0	273,6	1	3			1									1				
DK	NOTANK	NTK 55/11	55,00	11,0	578,9	1	3			1									1				
DK	WINMAT	WIMA 55/15	55,00	15,0	311,4	1	3			1									1				
DK	SONBJ	SM55	55,00	14,0	357,4	1	3			1									1				
DK	SONBJ	SONE 55/15	55,00	15,0	311,4	1	3			1									1				
DK	VEST/H	VEST 55/15	55,00	15,0	311,4	1	3	1		1									1				
DK	BONUS	BON 55/16.3	55,00	16,3	263,2	1	3	1	1	1			3,5	12,5	28,0	67,0	4	4	1		1	1	24,0
DK	BONUS	BON 55/18	55,00	18,0	216,3	1	3	1	1	1							4	4	1				
DK	VESTAS	VEST 55/16	55,00	16,0	273,6	1	3	1	1	1									1				
DK	MICON	M 55/11	55,00	16,0	279,3	1	3	1	1	1	1 NACA 44 modif.	5300	4,0	18,0	28,0	67,0	18	1	1		3	1	22,0
DK	REYMO	REY 55 kW	55,00	15,2	309,5	1	3			1	1								1		3	1	18,0
DK	WINCON	M 55/11	55,00	16,0	275,0	1	3	1	1	1			4,0		50,0	50,0	130	1	1		1	1	22,0
DK	WENERG	M 55/16	55,00	16,0	275,0	1	3	1	1	1			4,0	16,0	50,0	50,0	130	1	1		1	1	22,0
DK	FAST	Fasterholt 55kW	55,00	15,2	309,5	1	3			1	1												18,0
DK	MICON	M 300 - 55 kW	55,00	19,8	184,1	1	3	1	1	1	1			10,0			18		1	2	3	1	24,0
DK	WINMAT	WM 15 s	60,00	15,5	324,7	1	3	1	1	1	1 NACA 63-200		3,5	14,0	25,0	50,0	2	2	1	1	9	1	18,0
DK	MICON	M 60/13/US	60,00	16,0	304,7	1	3			1									1				24,0

Table 3 : WECs in the rated power range 10.1 - 75 kW (contd.)

CNT	MANUF.	TYPNAME	RATPOW kW	DIAM m	SPECPOW W/m²	ROTTYP #	NUMBLAD #	ORIENT #	POWCONT *	SPEED *	PROFILE #	WEIHEAD kg	VCUTIN m/sec	VRATED m/sec	VCUTOUT m/sec	VSURV m/sec	GENTYP *	TRANS *	BRKTYP *	BRKPOS #	TOWTYP *	TOWMAT *	TOWHEI m
DK	DANSME	SWE 65/16	65,00	16,0	323,4	1	3	1	1										1				
DK	NOTANK	NTK 65/16	65,00	16,0	323,4	1	3	1	1										1				
DK	VINCON	M66/13	65,00	16,0	330,1	1	3	1	1				4,0	17,0			4	4	1		1		
DK	VENERG	M 66/13-US	65,00	16,0	325,0	1	3	1	1										1				
DK	SMEMOE	SMEMOE 65/16.3	65,00	16,3	311,6	1	3	1	1										1				
DK	FAST	55/11	65,00	15,5	351,7	1	3	1	1								2		1				
DK	ALTERN	65/16	65,00	16,0	323,4	1	3	1	1										1				
DK	LM	LM 7.75	65,00	15,5	345,7	1	3	1	1										1				
DK	DANSME	SWE 75/20	75,00	20,0	243,7	1	3	1	1										1				
DK	NOTANK	NTK - 75F	75,00	16,6	347,2	1	3	1	1		1		3,5	19,0	23,0		18	1	1	2	3	1	22,0
DK	VINMAT	WIMA 75/17(V.2)	75,00	17,0	330,5	1	3	1	1										1		8	3	23,4
DK	VESTAS	VEST 75/17(V.2)	75,00	17,0	330,4	1	3	1	1		NACA 44xx	6700	4,0	15,0	23,0	50,0	4	1	1		3	1	22,5
DK	DENCON	D-75	75,00	17,8	307,8	1	30	1	1				4,0	13,0			4	4	1		8		18,0
DK	FAST	Fasterholt 75kW	75,00	17,0	337,4	1	3	1	1										1				24,0
DK	ALTERN	15-75	75,00	15,2	422,1	1	3	1	1									1	1		1	1	22,5
DK	WINFOS	75/17	75,00	17,0	330,4	1	3	1	1										1				
DK	ADWIP	Aero Tech 75	75,00	15,6	400,6	1	3	1	1		1								1				
DK	HANING	75 kW	75,00	17,0	337,4	1	3	1	1				4,0	18,0			4	4	1		1		22,5
DK	NORFOL	Folkecenter 75	75,00	17,0	337,4	1	3	1	1		1							1	1				23,0
E	GASE	PEUI-10/2	24,00	10,1	305,7	1	3	1	1		3 NACA	2000	4,0	11,0	25,0	65,0	2	1	1		8	1	12,0
E	ECOTEC	12/30	30,00	12,0	265,5	1	3	1	1		1		4,0	13,0	25,0	50,0	2	2		1	8		14,0
E	AERCAN	AC55	55,00	15,0	317,7	1	3	1	1				4,0	15,0			1	4	1	1			
F	RATIER	RF 70	70,00	15,0	404,4	1	3	1	1				5,0	19,0			2	2	1				
F	JELOU	LJ75A312	75,00	12,0	676,9	1	3	1	1		NACA 44xx												
FRG	BORN	6.6/11	11,00	6,6	321,6	1	3	1	1										1				
FRG	SCHUB	ES 10-11	11,00	10,0	142,9	1	1	1	1														
FRG	WIKZ	Elektromat 12	12,00	6,3	393,4	1	16	1	1									2	1				

Table 3 : WECs in the rated power range 10.1 - 75 kW (contd.)

CNT	MANUF.	TYPNAME	RATPOW kW	DIAM m	SPECPOW W/m²	ROTTYP #	NUMBLAD #	ORIENT #	POWCONT *	SPEED #	PROFILE	WEIHEAD kg	VCUTIN m/sec	VRATED m/sec	VCUTOUT m/sec	VSURV m/sec	GENTYP *	TRANS *	BRKTYP *	BRKPOS #	TOWTYP *	TOWMAT *	TOWHEI m
FRG	BRUEM	BW81	12,00	9,0	192,6	1	3	1	34				3,0	8,0			1	4			5	1	
FRG	SUEWIN	N 715	13,40	7,0	359,2	1	3	2	1	2			3,0		30,0	67,0	2	1	1	1	5	1	18,0 / 30,0
FRG	SUEWIN	N 718.5	15,00	7,0	397,9	1	3	2	1	2			3,0		30,0	67,0	2	1	1		5	1	18,0 / 30,0
FRG	BORN	8/15	15,00	8,0	300,0	1	3	1	1		NACA 44xx						1						
FRG	AERODY	Aeolus 11	18,00	11,7	167,4	1	3	2	18	1	NACA 4412-21		3,5	10,5	25,0		2	1	1		3	1	10,0 / 15,0
FRG	HSW	HSW-30	18,00	12,5	149,6	2	2	2	18	1	FX-W-84-140/218		4,5	8,9	20,0	53,0	2	1	1		1	1	14,5 / 22,3
FRG	SUEWIN	E 1220	20,00	12,5	166,3	1	3	2	1	1			3,0		30,0	67,0	1	1	1		5	1	18,5 / 30,5
FRG	WIKZ	Elektromat 20kW	20,00	10,5	230,9	1	3	1	1	4			3,5	11,0	30,0	67,0	17	1	1		8	1	14,0
FRG	MAN	Aeroman 12/20	20,00	12,0	177,0	1	2	1	18		NACA 4415-4424	1150	3,5	11,1		50,0	3	1	1		1	1	10,0 / 15,0
FRG	HMOT	HM-Rotor 20/60	20,00	10,0	333,3	2	3	3	1	4			4,0	12,5	30,0	75,0	128	16			5	1	
FRG	BRUEM	BW120	22,00	13,0	169,2	1	3	1	34				3,0	8,0			1	4			5	1	
FRG	SCHUB	ES 10-22	22,00	10,0	285,7	1	2	1	1								2						
FRG	SCHUB	ES 1000 L	22,00	10,0	285,7	1	2	1	1	4	NACA 44xx		3,0		20,0	50,0	1	1	1		5	1	
FRG	SUEWIN	E 1225	24,00	12,5	199,5	1	2	2	1	1			3,0		30,0	67,0	1	1	1		5	1	18,5 / 30,5
FRG	MBB	Monopterus 20	25,00	12,5	203,3	1	1	2	1				4,5	15,0	19,0	60,0	2		1		1	1	15,0 / 18,0
FRG	WIKZ	Elektromat 25kW	25,00	10,5	288,7	1	3	1	1	4			3,5	11,0	30,0	67,0	1	1	1		8	1	14,0
FRG	AERODY	25 KW aerodyn	25,00	12,5	203,3	1	2	2	18				4,5	10,9	20,0	53,0	2		1		1	1	14,5
FRG	HSW	HSW-30	25,00	12,5	207,8	1	2	2	18	1	FX-W-84-140/218						2	1	1		1	1	14,5 / 22,3
FRG	AERODY	Aeolus 12/220V	26,00	15,0	147,1	1	2	2	18	2	FX-77N	2520	2,0	12,0	25,0	53,0	2	1	1		5	1	18,0
FRG	KAEHL	Kano-Rotor 30kW	30,00	12,9	236,6	1	1	1	1	2	FX-W-151-A			12,0			18		1		3	1	15,5 / 24,5
FRG	SUEWIN	N 12/30	30,00	12,5	243,9	1	3	2	65	2		1700	3,0	12,0	30,0	67,0	2	1	1		5	1	24,0

Table 3 : WECs in the rated power range 10.1 - 75 kW (contd.)

CNT	MANUF.	TYPNAME	RATPOW kW	DIAM m	SPECPOW W/m²	ROTTYP #	NUMBLAD #	ORIENT *	POWCONT *	SPEED #	PROFILE	WEIHEAD kg	VCUTIN m/sec	VRATED m/sec	VCUTOUT m/sec	VSURV m/sec	GENTYP *	TRANS *	BRKTYP *	BRKPOS #	TOWTYP *	TOMAT *	TOWHEI m
FRG	DORN	Darrieus DZ-12	30,00	12,0	214,3	2	3	4	1		634-021 mod.		4,0	10,5	24,0	65,0	3	2	1				18,0
FRG	MAN	Aeroman 12.5/30	30,00	12,5	243,9	1	2	1		18	NACA 4415-4424	1290	3,7	11,8		50,0	4	1	1		1	1	10,0
FRG	BORN	10/30	30,00	10,0	384,6	1	3	1	1		NACA 44xx												15,0
FRG	BRUEM	BW160	30,00	15,0	173,3	1	3	1		34			3,0	8,0			1	4			5		
FRG	HUELL	F.H.W. 30	30,00	12,5	243,9	1	3	2	1	18	1	1500	3,5	12,0	20,0	60,0	2	1	1		3	1	15,0
FRG	SUEWIN	N 12/30	30,00	12,5	243,9	1	3	2	2	65	2	1700	3,0	12,0	30,0	67,0	2	1			5	1	
FRG	MBB	MBB 15/30	30,00	15,0	175,0	1	2	2													5	1	24,0
FRG	AERODY	Aeolus 12.5	33,00	12,5	274,5	1	2	1	1	18			4,6	12,4	25,0	55,0	2	4	1	1	1	1	20,0
FRG	MAN	Aeroman14.8/33n	33,00	14,8	191,9	1	2	1		18	1 NACA 4415-4424	1550	4,0	10,0			18	1	1	1	3	2	15,0 / 22,0
FRG	HSW	HSW-30	33,00	12,5	274,3	1	2	2	1		1 FX-W-84-140/218	1155	4,5	12,4	20,0	53,0	18	1	1	1	3	1	14,5 / 22,3
FRG	MAN	Aeroman14,8/33i	33,00	14,8	197,7	1	2	1		18	1 NACA 4415-4424	1550	4,0	10,0	25,0	55,0	17	1	1	1	3	2	15,0 / 22,0
FRG	MAN	Aeroman 12.5/40	40,00	12,5	332,8	1	2	1		18			4,0	11,8			4		1		1		
FRG	TACK	TW 45	45,00	12,5	366,7	1	3	3		1													
FRG	SUEWIN	N 1245	45,00	12,5	374,1	1	3	2	2	1	2		3,0	11,5	30,0	67,0	2	1	1	1	5	1	18,5 / 30,5
FRG	AERODY	Aeolus 12/380V	48,00	15,0	271,6	1	3	2		18	FX-77W	2520			30,0		2	2	1	1	5	1	18,0
FRG	KROG	125/50 G 30	50,00	12,9	394,3	1	3	1		1													
FRG	ENERC	ENERCON 16/55	55,00	16,3	268,3	1		1		1											3		22,0
FRG	ENERC	ENERCON 15/55	55,00	15,0	320,9	1	3	1		1											3	1	23,5
FRG	TACK	TW 60	60,00	16,9	504,2	1	3	3		1	1 Worthmann FX-84	3700	3,0	16,9	24,0	55,0	4	4	1		3	1	22,0
FRG	BSW	WR 65/17	65,00	17,0	286,3	1	3	1		9			3,0				4	2			3	2	29,3
FRG	RUHRTA	65 KW Ruhrtal	65,00	15,2	359,1	1	3	3	2								2	2	1		8	1	18,0
GR	WISO	BBW 20KDG	20,00	10,1	254,8	1	3	1		1	1 NACA 4415-28	1200	4,0	12,0	25,0	50,0	2	1	1		8	1	18,0

Table 3 : WECs in the rated power range 10.1 - 75 kW (contd.)

CNT	MANUF.	TYPNAME	RATPOW	DIAM	SPECPOW	ROTTYP	NUMBLAD	ORIENT	POWCONT	SPEED	PROFILE	WEIHEAD	VCUTIN	VRATED	VCUTOUT	VSURV	GENTYP	TRANS	BRKTYP	BRKPOS	TOWTYP	TOWMAT	TOWHEI
			kW	m	W/m²	#	#	#	*	#		kg	m/sec	m/sec	m/sec	m/sec	*	*	*	#	*	*	m
GR	VISO	BBW 30KDG	30,00	10,1	382,2	1	3	1	1	1	NACA 4415-28	1350	4,0	12,0	25,0	50,0	2	1	1		8	1	18,0
GR	HELLAI	Aiolos 55 kW	55,00	15,5	300,5	1	3	1	1	2	NACA 63-200	5400	3,0	14,0	27,0	50,0	17	4	1	1	3	1	24,0
																					3	1	30,0
																					10	3	24,0
																					10	1	30,0
I	AERIT	AIT 03	25,00	8,6	439,4	1	2	2	34				5,0	13,0			1				1	1	
I	RIVA	M 15 (25 kW)	25,00	17,0	113,5	1	1	2	2		FX 84-W series		4,8	8,5	16,0	60,0	1				1	1	
I	RIVA	M 15 (30 kW)	30,00	15,0	175,0	1	1	2	2		FX 84-W series		4,8	9,7	16,0	60,0	1				1	1	
I	TEMA	TEMA 2	40,00	14,0	265,3	1	4	1	1					11,0					1		1	1	
I	TEMA	TEMA 2	40,00	14,0	265,3	2	4		1								4						
IRL	WTI	Vanguard 95	65,00	15,5	351,7	1	3	1	18				4,0	14,0			4	4	1		8	1	21,0
NL	LAWEY	Enkel 11 kW	11,00	10,6	124,7	1	2	1	34								2		8		5	1	21,0
NL	LAWEY	Enkel 11 kW	11,00	10,6	124,7	1	2	1	34										8				
NL	FDO	6.3 WPX	14,00	6,3	448,7	1	1	1	1														
NL	POLENK	WPS10A15	15,00	9,6	207,2	1	1	1	1								2				1	1	17,0
NL	BOUMA	11m-22kW	15,00	11,0	161,1	1	1	1	1														
NL	GIRO	DVI 15/3	15,00	9,0	240,8	1	2	1	1														
NL	WYNFAN	11/15	15,00	11,0	157,9	1	1	1	1								2				5	1	21,0
NL	BERKOU	Windvang 80.15	15,00	8,0	304,9	1	3	1	34				4,0	12,0							8		
NL	WIPAOU	11/17.5	17,50	11,0	184,2	1	1	1	1								2				5	1	21,0
NL	DEJONG	10 J 18	17,50	10,3	210,1	1	1	1	1								2				1	1	49,0
NL	BERKOU	Windpaq	17,50	11,0	188,0	1	3	1	34				4,0	11,0			2				8	1	
NL	LAWEY	Twin	20,00	10,6	233,6	1	2	1	34								2		8		1		49,0
NL	H-ENSY	HE 1000	20,00	10,0	254,8	1	1	1	1														
NL	TCR	Aerotech 10	20,00	10,0	260,1	1	3	1	18				3,5	7,0			4		1		1	1	
NL	NEWINC	AEROTECH 10P120	20,00	9,1	314,0	1	3	1	18		NACA		4,0		25,0				1		9	5	
NL	LAWEY	Enkel 20 kW	20,00	10,6	226,5	1	3	1	1										8		5		
NL	BOUMA	11m-22kW	22,00	11,0	236,3	1	1	1	1														
NL	FDO	8 WPX	23,00	8,0	457,3	1	1	1	1														
NL	LAWEY	LW 24/10.6	24,00	10,6	272,1	1	3	1	34										8				

Table 3 : WECs in the rated power range 10.1 - 75 kW (contd.)

CNT	MANUF.	TYPNAME	RATPOW	DIAM	SPECPOW	ROTTYP	NUMBLAD	ORIENT	POWCONT	SPEED	PROFILE	WEIHEAD	VCUTIN	VRATED	VCUTOUT	VSURV	GENTYP	TRANS	BRKTYP	BRKPOS	TOMTYP	TOMMAT	TOWHEI
			kW	m	W/m²	#	#	#	*	#		kg	m/sec	m/sec	m/sec	m/sec	*	*	*	#	*	*	m
NL	LAWEY	LW 25 kW	25,00	15,0	144,4	1			34										8				
NL	LAWEY	Camperduin	30,00	10,6	347,2	1			34										8				
NL	BOUMA	11m-30kW	30,00	11,0	322,2	1			1														
NL	LAWEY	Enkel 35 kW	35,00	10,6	396,8	1	3	1	34				3,0	12,0			2	4	8		1	1	24,0
NL	LAWEY	LW 11/35	35,00	10,6	397,7	3	3	1	34				3,0	12,0			16	1	8		1	1	24,0
NL	FDO	10 WPX	36,00	10,0	458,6	1			1										1				
NL	GBO	WG 12	37,00	12,5	307,8	1	3	1	1				4,0	12,0			4		1				
NL	H-ENSY	HE 1500 L	37,00	15,0	213,7	1	2	1	4				4,5	12,0			2	4	1		1	1	17,0
NL	POLENK	WPS11A40	40,00	11,5	385,0	1		2	1								2		1				
NL	MULTI	45 kW	45,00	16,0	228,5	1	2	2	1								4		1				
NL	TCR	Aerotech 14	50,00	14,0	331,6	1	3	1	18				3,5	7,0	25,0				1		1	1	
NL	NEWINC	AEROTECH 14PS50	50,00	13,8	341,3	1	3	1	18		NACA		4,0			25,0			1		9	1	24,0
NL	LAWEY	LW 50/15,6	50,00	15,6	268,8	1			34										8				20,0
NL	POLENK	(55/16)	55,00	16,0	279,3	1	3	1	1				3,7		20,0	56,0	128	1	1				24,0
NL	GBO	WG 16	55,00	16,0	279,3	1	2	2	1								4		1				
NL	BOUMA	16m-55kW	55,00	16,0	273,5	1		1	1								2		1		1	1	17,0
NL	H-ENSY	HE 1600	55,00	16,0	279,3	1	3	1	2				4,5	12,0			2	4	1		1	1	
NL	H-ENSY	HE 2000	55,00	20,0	178,7	1		1	1										1				
NL	FDO	12.5 WPX	57,00	12,5	464,5	1			1														
NL	POLENK	WPS16A60	60,00	16,0	298,4	1		1	1								2		1		1	1	17,0
NL	POLENK	WPS16SM60	60,00	16,0	298,4	1		1	1								1		1		1	1	17,0
NL	KAAL	Single	60,00	15,0	346,6	1			1										1				
NL	BERMOU	Windvang 160.6	60,00	16,0	304,7	1		2	2				4,0	12,0			2	2	1				
NL	NEWINC	AERO 14P1 50/65	65,00	13,8	443,7	1	3	1	18		NACA		4,0	12,0		25,0			1		9	1	24,0
NL	LAWEY	LW 15/75	75,00	15,6	404,5	1	2	1	34				3,0	13,0			16		8		3	1	30,0 / 36,0 / 42,0
NL	LAWEY	sixmaster	75,00	15,2	426,1	1	2	2	34				3,0	13,0			16	1	8		1	1	55,8
NL	GBO	KEWT 1	75,00	16,0	380,9	1	2	2	1								4		1				
NL	BOUMA	16m-75kw	75,00	16,0	372,9	1		1	1										1				

Table 3 : WECs in the rated power range 10.1 - 75 kW (contd.)

CNT	MANUF.	TYPNAME	RATPOW	DIAM	SPECPOW	ROTTYP	NUMBLAD	ORIENT	POWCONT	SPEED	PROFILE	WEIHEAD	VCUTIN	VRATED	VCUTOUT	VSURV	GENTYP	TRANS	BRKTYP	BRKPOS	TOMTYP	TOMMAT	TOWHEI
			kW	m	W/m²	#	#	*	#	#		kg	m/sec	m/sec	m/sec	m/sec	*	*	*	#	*	*	m
NL	LAWEY	Quadro	75,00	15,6	98,1		2		1											8	3	1	
UK	MMEW	9/15 kW	15,00	9,0	240,8	2	3		1				2,0					4			5		
UK	AEOLTD	7M	22,00	7,0	583,6	1	3	2	34								256	4			8	1	
UK	NEI	25kW	25,00	10,8	278,7	1	3		1				4,5	11,0			64	4					
UK	GRYLLS	WL40LC	40,00	9,5	571,4	1	3		1	4	NACA 0018						8	9	1		5	1	9,5
UK	LAWSON	Windiesel	55,00	17,5	234,0	1	3	1	99	4	NACA 0025	3200	7,0	13,0	22,0	50,0	1	32	1		8	1	12,0
UK	HOWDEN	HWP-60	60,00	15,0	346,6	1	3	2	1			2700	4,0	15,0			5	1	1		1	1	20,0
UK	NEI	60kW	60,00	15,2	337,6	1	3		1				5,0	12,0			64	4	1				
J	KIDEN	SW-VAWT	45,00	8,0	912,8	2	3		1	4			3,5	15,0	15,0				1				13,8

Table 4 : WECs in the rated power range 75.1 - 300 kW

CNT	MANUF.	TYPNAME	RATPOW kW	DIAM m	SPECPOW W/m²	ROTTYP #	NUMBLAD #	ORIENT #	POWCONT *	SPEED #	PROFILE	WEIHEAD kg	VCUTIN m/sec	VRATED m/sec	VCUTOUT m/sec	VSURV m/sec	GENTYP *	TRANS *	BRKTYP *	BRKPOS #	TOMTYP *	TOMMAT *	TOWHEI m
B	HMZ	WindMaster 100	100,00	21,8	273,5	1	3	1	18	1	NACA 4418-4437	9000	4,0	14,0	24,0	65,0	2	1	1	2	1	1	17,0
																						1	22,0
																						1	30,0
B	HMZ	WindMaster 150	150,00	21,8	402,1	1	3	1	18	1	NACA 4418-4437	9100	4,5	15,0	24,0	65,0	18	1	3		3	1	22,0
																					3	1	30,0
B	HMZ	WindMaster 175	175,00	21,8	478,7	1	3	1	50	1	NACA 643-618		5,0	14,5	25,0		130		1		1	1	22,0
B	HMZ	WindMaster 200	200,00	22,5	518,5	1	3	1	18	1	NACA 4418-4437	9200	5,0	15,0	24,0	65,0	18	4	3		3	1	22,1
																						1	29,1
																						1	42,0
B	HMZ	WindMaster 225	225,00	22,5	577,7	1	3	1	50	1			5,0	13,0			4	4	1		1	1	30,0
B	HMZ	WindMaster 250	250,00	25,0	510,2	1	3	1	18	1	NACA 4418-4437	9650	5,0	14,0	24,0	65,0	18	4	3		3	1	22,0
																					3	1	53,0
B	HMZ	WindMaster 300	300,00	25,0	624,0	1	3	1	18	1	NACA 4418-4437	12000	5,0	15,5	24,0	65,0	18	4	3		3	1	21,8
																					3	1	30,0
																					3	1	53,0
DK	DANSME	SME 80/18	80,00	18,0	314,6	1	3	1	1										1			1	18,0
DK	WINMAT	wm 17 s	80,00	17,0	359,9	1	3	1	1		NACA 63-200		3,5	14,0	25,0	50,0	2	1	1		9	1	22,0
																						1	24,0
DK	DWT	T-17/80	80,00	17,0	359,9	1	3	1	1				3,0	14,0			4	4	1		3	1	
DK	DWT	WINDANE 17/80	80,00	17,0	352,4	1	3	1	1	2	NACA 63-200	5380	3,0	14,0	28,0	65,0	18	4	1	1	3	1	22,6
																					10	1	24,6
																						1	28,5
																					3	1	32,4
																						1	40,2
DK	TELL	UK T-1780	80,00	17,0	352,4	1	3	1	1										1		10	1	
DK	SMEMOE	SMEMOE 80/18.7	80,00	18,7	291,4	1	3	1	1														
DK	SVEDAN	Svedana 80/50Hz	80,00	17,0	359,9	1	3	1	1		NACA 63-200		3,5	15,0	27,0	50,0	2	1	1		9	1	24,0
DK	DVTNAK	80kW	80,00	20,6	245,0	1	3	1	1				4,0	16,3			4	4	1		8	1	
DK	DVTNAK	Smedemoellen 80	80,00	18,7	297,4	1	3	1	1	1									1			1	24,0

Table 4 : WECs in the rated power range 75.1 - 300 kW (contd.)

CNT	MANUF.	TYPNAME	RATPOW kW	DIAM m	SPECPOW W/m²	ROTTYP #	NUMBLAD #	ORIENT #	POWCONT *	SPEED #	PROFILE	WEIHEAD kg	VCUTIN m/sec	VRATED m/sec	VCUTOUT m/sec	VSURV m/sec	GENTYP *	TRANS *	BRKTYP *	BRKPOS #	TOMTYP *	TOMMAT *	TOMHEI m
DK	AIOLA		80,00	17,0	363,3														1				24,0
DK	DAVID	Lolland 80/18m	80,00	18,7	300,3	1	3		1										1		10	1	24,0
DK	DANSME	SME 90/20	90,00	20,0	286,6	1	3		1										1		3	1	30,0
DK	VESTAS	V19-90	90,00	18,8	324,9	1	3		1		NACA 44xx	7000	3,4	18,0	23,0	50,0	4	1	1		8	1	23,4
DK	VESTAS	VEST 90/18	90,00	18,0	353,9	1	3		1	1									1		3	1	22,5
DK	VESTAS	V 20	90,00	20,0	286,6	1	3		1				4,0	15,0	25,0				1		1	1	24,0
DK	DANWIN	Danwin 19	90,00	19,6	304,6	1	3		1	1									1		1	1	
DK	FAST	75/15	90,00	17,0	404,9	1	3		1			3700	4,0	15,5	30,0		2	2	1		1	1	22,5
DK	ALTERN	15-90	90,00	15,6	480,8	1	3		1										1		1	1	22,5
DK	LM	LM 8.5	90,00	17,0	396,5	1	3		1										1		3	1	28,5
DK	ADWIP	AWP 90/18	90,00	18,5	345,2	1	3		1				4,0	15,5			18	4	1	1	3	1	28,5
DK	VISYS	AWP 90/18	90,00	18,0	364,7	1	3		1										1		10	1	28,5
DK	WINCON	M110 (50Hz)	93,00	19,0	334,9	1	3		1				4,0	15,0			4	4	1		3	1	30,0
DK	WINCON	W110/XT (50Hz)	93,00	21,0	274,1	1	3		1	1			4,0	15,0			18	4	1	2	3	1	30,0
DK	WINMAT	WM 19S	95,00	19,0	342,1	1	3		1				3,5	13,0			4	4	1		8	1	22,0
DK	BONUS	BON 95/19.4	95,00	19,4	321,6	1	3		1								18	4	1		3	1	27,0
DK	MICON	M 95kW	95,00	19,3	331,5	1	3		1	1							18		1		34	1	30,0
DK	REYMO	REY 20/95kW	95,00	19,6	324,6	1	3		1								18		1		10	1	24,6
DK	DWT	WINDANE 19/95	95,00	19,0	334,5	1	3		1	2	NACA 63-200	6015	3,0	14,0	28,0	65,0	18	4	1		10	1	32,4

Table 4 : WECs in the rated power range 75.1 - 300 kW (contd.)

CNT	MANUF.	TYPNAME	RATPOW (kW)	DIAM (m)	SPECPOW (W/m²)	ROTTYP (#)	NUMBLAD (#)	ORIENT (*)	POWCONT (*)	SPEED (#)	PROFILE	WEIHEAD (kg)	VCUTIN (m/sec)	VRATED (m/sec)	VCUTOUT (m/sec)	VSURV (m/sec)	GENTYP (*)	TRANS (*)	BRKTYP (*)	BRKPOS (#)	TOMTYP (*)	TOMMAT (*)	TOMHEI (m)
DK	DWT	VINDANE 19/95	95,00	19,0	334,5	1	3	1	1	4		6015					16	1	1		10	1	40,2
DK	TELL	Tellus/DWT 95	95,00	19,0	341,8	1	3		1									1	1				
DK	WINCON	WINC 95/19.6	95,00	19,6	315,0	1	3		1									1	1				
DK	WENERG	M 95	95,00	19,3	324,2	1	3	1	1		NACA 63-200		4,0	16,0	28,0	50,0	130	1	1		1	1	22,0
DK	SVEDAN	Svedana100/60Hz	95,00	17,0	427,4	1	3		1				3,5	15,0	27,0	50,0		1	1		9	1	24,0
DK	DANSME	SME 99/21.3	99,00	21,3	278,0	1	3		1										1				
DK	NOTANK	NTK 99/22	99,00	22,0	260,6	1	3		1										1				
DK	NOTANK	NTK 99F	99,00	20,5	300,1	1	3		1										1		3	1	24,5
DK	WINCON	WINC 99/20	99,00	20,0	315,3	1	3		1										1				
DK	SOEG	99kW/19.6m	99,00	19,6	335,0		3		1										1				24,0
DK	WINMAT	WIMA 99/19	99,00	19,0	360,0	1			1										1				
DK	VESTAS	VESTAS 99/20	99,00	20,0	324,9		3		1										1				
DK	NOTANK	NTK 100/18.2	100,00	18,2	384,6	1	3		1										1		34		
DK	BONUS	BON 100/20	100,00	19,4	348,8	1	3		1	2	NACA 44 modif.	8400	3,5	12,5	28,0	67,0	18	4	1	1	3	1	30,0
DK	MICON	M 100	100,00	19,0	360,1	1	3		1	1	NACA 44 modif.		4,0	18,0	28,0	67,0	18	1	1		3	1	24,0
DK	DWT	T-17/100	100,00	17,0	449,8	1	3		1				3,0	15,0			4	4	1		3	1	22,0
DK	WINCON	M 100 (V.2)	100,00	19,0	352,1	1	3		1				4,0	13,5	45,0		130	1	1		1	1	22,0
DK	WIWOR	Wind-World 1	100,00	19,5	341,9	1	3		1										1				
DK	BALPOW	100kW	100,00	19,6	338,4	1	3		1				4,2	15,0			4	4	1		8	1	23,4
DK	VESTAS	V 20-100kw	100,00	20,0	318,5	1	3		1	1	NACA 44xx	6300	4,5	13,0	25,0		18		1	1	3	1	23,2
DK	MICON	M 100/US	108,00	19,0	388,9	1	3		1				4,0	15,0			4	4	1		1	1	
DK	WINCON	M110 (60Hz)	108,00	19,0	388,9	1	3		1				4,0	15,0			4	4	1		1	1	
DK	WINCON	W110/XT (60Hz)	108,00	21,0	318,3	1	3		1								4	4	1				
DK	WENERG	M 108	108,00	19,3	368,6	1	3		1								4	4	1				
DK	DWP	D 110	110,00	19,2	387,9	1	3		1				4,0	14,0			4	4	1		1	1	
DK	DANWIN	19/110	110,00	19,0	396,1	1	3		1				3,5	11,0			2	4	1		1	1	
DK	WIWOR	W-1960	110,00	19,6	364,2	1	3		1		NACA 4412-26	6400	4,0	14,0	25,0	67,0	4	1	1		1	1	
DK	DWTNAK	80kW	110,00	20,6	336,9	1	3		1				4,0	16,3			4	4	1		8	1	23,0

Table 4 : WECs in the rated power range 75.1 - 300 kW (contd.)

CNT	MANUF.	TYPNAME	RATPOW kW	DIAM m	SPECPOW W/m²	ROTTYP *	NUMBLAD #	ORIENT *	POWCONT #	SPEED *	PROFILE #	WEIHEAD kg	VCUTIN m/sec	VRATED m/sec	VCUTOUT m/sec	VSURV m/sec	GENTYP *	TRANS *	BRKTYP *	BRKPOS #	TOWTYP *	TOMMAT *	TOWHEI m
DK	DANWIN	19/120	120,00	19,0	432,1	1	3	1	1				3,5	11,0			2	4	1		1	1	
DK	WIMOR	Wind-World 120	120,00	20,2	382,0	1	3	1	1										1				
DK	NOTANK	NTK - 130F	130,00	20,5	393,9	1	3	1	1		1						18	3	1	2	3	1	24,0
DK	WINMA	WIMA 130	130,00	19,4	448,7	1	3	1	1										1				
DK	DWT	WINDANE 19/130	130,00	19,0	457,7	1	3	1	1		4	6015					16		1		9	1	22,6 24,6 40,2
DK	VISYS	v-s 130	130,00	22,0	342,2	1	3	1	1										1				
DK	WIMOR	W-2250 (50Hz)	130,00	22,5	333,8	1	3	1	1				4,0	13,0			4	4	1		1	1	
DK	NORFOL	Folkec. 130 s	130,00	22,0	349,2	1	3		1		1								1				
DK	WINMA	WIMA 130/20	130,00	20,0	426,6	1	3	1	1										1				
DK	DWP	D 140	140,00	19,6	463,6	1	3	1	1										1				
DK	DANWIN	Darwin 19/140	140,00	19,6	463,6	1	3	1	1		NACA 44xx		3,5	14,0	25,0	67,0	2	1	1		3	1	22,5
DK	DANWIN	23/140	140,00	23,2	338,1	1	3	1	1				3,5	11,0			2	4	1		1	1	
DK	NOTANK	NTK150	150,00	20,5	464,0	1	3	1	1				3,7	18,0			2	2	1				
DK	NOTANK	NTK 150 XLR	150,00	24,6	315,8	1	3	1	1		1 NACA 63xx		4,0	13,0	25,0	53,0	18	1	1		3	1	30,0
DK	WINMA	WM 22S	150,00	22,0	402,9	1	3	1	1				3,0	13,0			4	4	1		8	1	
DK	BONUS	Bonus 150 kW	150,00	23,0	361,4	1	3	1	1		2 NACA 63-200	11000	4,0	14,0	28,0	67,0	18	4	1	1	10	1	
																					34	1	24,0
																					3	1	30,0
																					3	1	26,0
																					32	1	30,0
DK	REYMO	REY 30/150kW	150,00	22,0	402,9	1	3	1	1								18	1	1		10	1	24,0
DK	WINCON	W 150 XT	150,00	24,0	331,9	1	3	1	1		1		4,0	13,0	25,0	52,0	2		1		1	1	30,0
DK	VISYS	V-S 150 (V.2)	150,00	22,0	402,9	1	3	1	1		1 NACA 63 modif.		3,7	14,0	25,0		18		1	2	10	1	30,0
DK	DWP	D 150	150,00	22,2	387,6	1	3	1	1			7100	3,0		25,0	60,0	18	1	1		3	1	23,5
DK	WIMOR	W-2250 (60Hz)	150,00	22,5	385,1	1	3	1	1				4,0	13,0			4	4	1		3	1	29,5

Table 4 : WECs in the rated power range 75.1 - 300 kW (contd.)

CNT	MANUF.	TYPNAME	RATPOW	DIAM	SPECPOW	ROTTYP	NUMBLAD	ORIENT	POWCONT	SPEED	PROFILE	WEIHEAD	VCUTIN	VRATED	VCUTOUT	VSURV	GENTYP	TRANS	BRKTYP	BRKPOS	TOWTYP	TOWMAT	TOWHEI
			kW	m	W/m²	#	#	*	#	*		kg	m/sec	m/sec	m/sec	m/sec	*	*	*	#	*	*	m
DK	WIWOR	Wind-World 150	150,00	23,2	362,1	1	3	1	1		NACA 4412-4430						18		1		3	1	31,0
DK	ALTERN	150/19.6	150,00	19,6	496,7	1	3	1	1										1				
DK		150 kW	150,00	22,0	402,9	1	3	1	1				4,2	14,0			4		1		8	1	30,0
DK	NORDEX	Nordex 150 kW	150,00	27,0	270,1	1	3	1	1	2			3,0	10,0	25,0		18	1	1		3	1	30,0
DK	MOEN		150,00	22,0	387,9																10	1	22,0
DK	DANWIN	Danwin 24/150	150,00	24,0	341,8	1	3	1	1								18		1		3	1	29,0
DK	MICON	M 450-150 kW	150,00	24,0	341,8	1	3	1	1										1		3	1	
DK	WIWOR	W 150/25m	150,00	25,0	315,0	1	3	1	1										1		3	1	31,0
DK	WIWOR	W 150/27m	150,00	27,0	270,1	1	3	1	1										1		3	1	31,0
DK	VISYS	V-S 150 (V.3)	150,00	23,0	372,2	1	3	1	1	1	NACA 63 modif.		3,7	14,0	25,0		18	1			3	1	
DK	VESTAS	VESTAS 150 kW	150,00	25,0	315,0	1	3	1	1											2			
DK	WIWOR	W-2800 150kW	150,00	28,0	243,5	1	3	2	1	1	NACA 63-??	13000	4,0	12,0	25,0				1	2	3	1	30,0
DK	NORDEX	Nordex 150/26	150,00	26,0	291,3	1	3	1	1										1				
DK	DANWIN	Danwin 25	160,00	25,5	319,9	1	3	1	1	1									1				24,0
DK	WIWOR	Wind-World 160	160,00	23,5	376,4	1	3	1	1										1				31,0
DK	LM	LM 11	160,00	23,0	385,5	1	3	1	1												3	1	
DK	DWP	D 175L	175,00	22,2	452,2	1	3	1	1	1		7900	3,0		25,0	60,0	18		1		3; 3	1	23,5; 29,5
DK	DWP	D 175/24.6	175,00	24,6	379,6	1	3	1	1										1		3		
DK	DWP	D 175/23.6	175,00	23,6	412,4	1	3	1	1										1				
DK	WINMAT	WIMA 22/32	180,00	23,0	442,0	1	3	1	1										1				
DK	VISYS	V-S 180	180,00	22,2	465,1	1	3	1	1				4,0	14,0	25,0		2	1	1		9	1	26,0
DK	DANWIN	Danwin 23/180	180,00	23,2	434,7	1	3	1	1				3,5	11,0			2	4	1		3	1	29,0
DK	WIWOR	Wind-World 180	180,00	23,2	438,9	1	3	1	1										1		3	1	31,0
DK	BONUS	BON 190/19.4	190,00	19,4	643,2	1	3	1	1								4	4	1		1	1	
DK	WINCON	WINC 198/21	198,00	21,0	571,9	1	3	1	1														
DK	WINMAT	WIMA 200	200,00	23,0	491,2	1	3	1															
DK	VESTAS	VESTAS 25-200	200,00	25,0	407,3	1	3	1	1	2	NACA 44xx	10300	3,6	13,8	25,0	56,0	18	1	2	1	3; 8	1	28,7; 29,0

Table 4 : WECs in the rated power range 75.1 - 300 kW (contd.)

CNT	MANUF. TYPNAME	RATPOW kW	DIAM m	SPECPOW W/m²	ROTTYP #	NUMBLAD #	ORIENT	POWCONT #	SPEED	PROFILE #	WEIHEAD kg	VCUTIN m/sec	VRATED m/sec	VCUTOUT m/sec	VSURV m/sec	GENTYP	TRANS	BRKTYP	BRKPOS #	TOMTYP #	TOMMAT	TOWHEI m
DK	WINCON W 200	200,00	24,0	455,8	1	3	3	1	1			4,5	14,0	25,0	52,0	18	4	1	2	3	1	28,7
DK	DENCON D-200	200,00	25,8	390,5	1	3	3	1				4,0	13,0				4			8		
DK	DANWIN Danwin 24	200,00	24,0	442,5	1	3	3	1	1	NACA 63-200		4,0	14,0	25,0	67,0	18	1		2	3	1	29,0
DK	DANWIN Danwin 24 (US)	200,00	24,0	442,5	1	3	3	1	1	NACA 63-200		4,0	14,0	25,0	67,0	2	1			1	1	29,0
DK	WIMOR W-2300	200,00	23,0	481,9	1	3	3	1		NACA 4412-34	7400	4,0	13,0	25,0	67,0	4	1			1	1	23,0 / 29,0
DK	WIMOR 200kW/415m²2	200,00	23,0	481,9	1	3	3	18		NACA 44xx		3,5		25,0	56,0			1		1	1	23,6
DK	WIMOR Wind-World 2	200,00	23,5	470,8	1	3	3	1														
DK	DAVID Lolland	200,00	25,5	403,7	1	3	3	1														
DK	ALTERN 200/23.3	200,00	23,0	469,5	1	3	3	1														
DK	ADWIP AWP 200/23	200,00	23,0	481,9	1	3	3	1	1			4,0	15,0	25,0	57,0	4				3		30,0
DK	FLEMA 200kW	200,00	25,2	409,4	1	3	3	18														
DK	CO-DAN Co-Dan 200 kW	200,00	25,5	403,7	1	3	3	1												3		30,0
DK	WIMOR W-2500 220kW	220,00	25,0	448,1	1	3	3	1		NACA 63-77		5,0	14,0	25,0		4			2	3	1	30,0
DK	VISYS V-S 225	225,00	24,0	497,8	1	3	3	1				4,0	14,0	25,0		2	1			10	1	26,0
DK	NORDEX Nordex 225/250	225,00	26,0	436,9	1	3	3	1	2			3,5	13,0	25,0		18	1			9	1	30,0
DK	VESTAS V-27/225	225,00	27,0	405,1	1	3	3	18	2	NACA 44xx						18	1		1	10	1	30,0
DK	DANWIN Danwin 27/225	225,00	27,0	405,1	1	3	3	1		NACA 63-200						18	1	2		3	1	
DK	WENERG M24 (250kW)	250,00	24,0	564,2	1	3	3	1				4,0	14,0			4	4	4		1		29,0
DK	DENCON D-250	250,00	20,2	795,9	1	3	3	1														
DK	DENCON D-250	250,00	25,5	499,5	1	3	3	1												3		30,0
DK	MICON M 450-250kW	250,00	24,0	553,1	1	3	3	1	1		11500	4,0	14,5	28,0		18		1	2	3	1	28,7
DK	MICON M 530-250kW	250,00	26,0	471,7	1	3	3	1	1		12100	4,0	13,0	25,0	67,0	18	1	1	2	3	1	28,7
DK	DANWIN Danwin 25/250	250,00	25,0	525,0	1	3	3	1												3		29,0
DK	DWT DWT 30/265	265,00	30,8	363,1	1	3	3	1												3		24,0
DK	VISYS V-S 270	270,00	26,0	508,5	1	3	3	1	1	NACA 63		4,0	16,0	24,0		18	1			3		30,0
DK	ALTERN 275/25.3	275,00	25,3	546,7	1	3	3	1												3		40,0

Table 4 : WECs in the rated power range 75.1 - 300 kW (contd.)

CNT	MANUF.	TYPNAME	RATPOW kW	DIAM m	SPECPOW W/m²	ROTTYP #	NUMBLAD #	ORIENT #	POWCONT *	SPEED #	PROFILE	WEIHEAD kg	VCUTIN m/sec	VRATED m/sec	VCUTOUT m/sec	VSURV m/sec	GENTYP *	TRANS *	BRKTYP *	BRKPOS #	TOMTYP *	TOMMAT *	TOWHEI m
DK	REYMO	REY 275	275,00	26,0	534,0	1	3	1	1										1		3	1	40,0
DK	NOTANK	NTK 300	300,00	27,5	505,1	1	3	1	1	1	NACA 63-200		4,5	13,0	25,0	53,0	18	1	1		3	3	
DK	DWT	WINDANE 31/300	300,00	30,8	402,7	1	3	2	18		NACA 4412	24200	6,0		25,0	67,0	132	1	1		1	1	30,8
DK	BONUS	BONUS 300 kW	300,00	35,0	321,4	1	3	1	1														
DK	NOTANK	NTK 300/31	300,00	31,0	409,8	1	3	1	1														
E	AMPP	56-100	100,00	18,0	405,2	1	3	2	2												10	1	18,0
E	ECOTEC	20/150	150,00	20,0	477,7	1	3	1	1	2			4,0	14,0	25,0	50,0	18	1	1	2	3	1	29,0
E	CENEME	F-19	300,00	19,0	1080,3	2	2	4	1				6,0	20,0			1	4			10	1	24,0
F	RATIER	RF 100	100,00	18,0	401,1	1	2	1	1		HOR		5,0	12,0	25,0	60,0	130	16	1		1	1	30,0
FRG	BSW	WR80/18	80,00	18,0	314,3	1	3	1	9				3,5	13,0			4	2			8	1	
FRG	ENERC	ENERCON 17/80	80,00	17,2	354,9	1	3	1	16	4	NACA 4415,4424		2,5	13,5	24,5		1	4	2	2	3	2	28,0
FRG	NOAH	Noah90kW	90,00	17,0	418,6	1			1														
FRG	WIKZ	Elektromat 90	90,00	18,6	333,3	1	3		1	1			4,0	12,5	25,0				1		8	1	24,0
FRG	KOEST	Adler 25/100kW	100,00	25,0	208,0	1	3	2	18	2	FX-77W-xxx	10260	3,4	10,5	20,0	60,0	18	1	2	2	5	1	22,0
FRG	PIBAM	WR 100	100,00	20,2	318,5	1		1	1														
FRG	NOAH	NOAHGrundmodell	100,00	17,0	440,5	1	5	1	1				3,0	12,3			3	1					
FRG	SCHUB	ES 2000 L	100,00	20,0	324,8	1	2	1	1	4	NACA 4412		3,0		25,0	50,0	2				3	1	24,0
FRG	VENTIS	20-100	100,00	20,0	328,2	1	2	1	34	2	NACA 44xx		3,3	10,7		60,0	2		1		3	1	30,0
FRG	TACK	WR 150	150,00	23,4	348,8	1	3	1	1	2	NACA 44												
FRG	TACK	TW 150	150,00	20,5	454,4	1	3	1	1														
FRG	WIKZ	Elektromat 150/27	150,00/27	27,0	270,1	1	3	1	1	4	NACA 3600		3,5	12,5	25,0	60,0	2			2	10	1	30,0
FRG	TACK	TW 150	160,00	21,0	476,2	1	3	1	1	1	NACA 63-200		4,0	14,5	24,0	60,0		1	1		3	2	28,5
FRG	KOEST	Adler 25/165kW	165,00	25,0	346,5	1	3	2	18	2	FX-77Wx,NACA44	10260	3,4	13,5	20,0	60,0	18	1	1	2	5	2	22,0

Table 4 : WECs in the rated power range 75.1 - 300 kW (contd.)

CNT	MANUF.	TYPNAME	RATPOW kW	DIAM m	SPECPOW W/m²	ROTTYP #	NUMBLAD #	ORIENT	POWCONT	SPEED	PROFILE #	WEIHEAD kg	VCUTIN m/sec	VRATED m/sec	VCUTOUT m/sec	VSURV m/sec	GENTYP	TRANS	BRKTYP	BRKPOS #	TOWTYP	TOMMAT	TOWHEI m
FRG	MBB	Monopterus 30	200,00	30,0	288,9	1	1	2	18								1		1		1	1	
FRG	HSW	HSW-250	250,00	25,0	510,2	1	3	1	1		1 NACA 4424-18	12000	4,0	13,0	23,0		18	2	1	1	3	1	27,3
FRG	TACK	TW 250	250,00	24,0	569,7		3	1	1		1 NACA 63-200		4,0	16,0	24,0	60,0			1		3	2	28,5
FRG	VIKZ	Elektromat250/27	250,00	27,0	450,1	1	3	1	1		4 NACA 3600		3,5	12,5	25,0	60,0	2	1	1	2	10	1	30,0
FRG	ENERC	ENERCON 32/280	280,00	32,0	358,9	1	3	1	18		4 NACA 4424-4415		3,0	11,3	25,0		18	2	1	2	3	2	34,5
FRG	ENERC	ENERCON 32/300	300,00	32,0	373,1	1	3	1	18		4 NACA 4424-4415		3,0	11,5	25,0	70,5	1	2	1	2	3	2	34,5
FRG	HMOT	HM-Rotor 300	300,00	32,0	446,4	2	2	3	1		4		4,0	13,5	28,0		128	16			5	1	
GR	HELLAI	Aiolos 100 kW	100,00	19,0	363,6	1	1	1	1		2 NACA 63-200	6900	3,0	15,0	27,0	50,0	17	4	1	1	3, 3, 10, 10	1	24,0 / 30,0 / 24,0 / 30,0
I	RIVA	M 30	200,00	33,0	241,1	1	1	2	18		4 FX 84-W-127		4,5	11,0	20,0	55,0	2		1			1	30,0
I	AERIT	AIT 02 - Medit	225,00	32,0	279,9	1	2	1	2		1 NACA 4412/20		4,2	9,8	20,0	50,0	4		1				25,0
IRL	WTI	Vanguard 95KW	95,00	17,0	427,4	1	3	1	18		2		3,5	15,0	25,0	50,0	130	25	1		8	1	25,0
NL	BERKOU	Windvang 160.80	80,00	16,0	406,3	1	2	2	2				5,0	13,0	13,0		2	2	1				
NL	TCR	Aerotech 17	85,00	17,0	382,4	1	3	1	18				3,5	7,0			4		1		1	1	
NL	NEWINC	AEROTECH 17PS85	85,00	17,0	382,4	1	3	1	18		NACA		4,0		25,0				1		9	1	
NL	FDO	16 WPX	90,00	16,0	447,5	1			1										1				
NL	POLENK	WPS18A100	100,00	18,0	392,9	1			1								2				1	1	17,0
NL	BOHEM	NBK 100	100,00	17,0	449,8	1	2	1	2				5,0	13,0			1		1		1	1	
NL		VAWT 15	100,00	15,0	666,7	1			1												1	1	17,0
NL	NEWINC	AERO 17P185/100	100,00	17,0	449,8	1	3	1	18		NACA		4,0		25,0				1		9	1	
NL	H-ENSY	HE 3000	110,00	30,0	158,9	1		3	1										1				
NL	KAAL	Twinmaster	120,00	16,0	298,4	1	3		1								1		1		8	1	9,0
NL	TRASCO	TWS 125	125,00	22,0	335,8	1			1								2		1				
NL	FDO	20 WPX	145,00	20,0	461,5	1			1														

Table 4 : WECs in the rated power range 75.1 - 300 kW (contd.)

CNT	MANUF.	TYPNAME	RATPOW	DIAM	SPECPOW	ROTTYP	NUMBLAD	ORIENT	POWCONT	SPEED	PROFILE	WEIHEAD	VCUTIN	VRATED	VCUTOUT	VSURV	GENTYP	TRANS	BRKTYP	BRKPOS	TOWTYP	TOMMAT	TOWHEI
			kW	m	W/m²	#	#	#	*	#	#	kg	m/sec	m/sec	m/sec	m/sec	*	*	*	#	*	*	m
NL	BERWOU	Windvang 200.15	150,00	22,0	402,9	1	2	1	1	18			4,5	13,5			1	2	1			1	
NL	BOUMA	Bouma 160/20	160,00	20,0	520,0	2	3	1	1	1			5,0	12,0			2	4	1			1	
NL	POLYMA	Pionier II	160,00	21,0	471,6	2	2	1	1	1			4,5	18,0			2	2	1		1	1	
NL	BOUMA	24.5m/250kW	250,00	24,5	541,4	1	3	1	1	18	1 NACA series		5,0		30,0	45,0	18	4	1		1	1	25,0
NL	TCR	Aerotech 23	250,00	23,1	609,0	1	3	1	1	18			3,5	7,0			4					1	30,0
NL	NEWINC	AER023P1200/250	250,00	23,1	609,0	1	3	1	1	18	NACA		4,0		25,0				1			1	
NL	POLENK	WPS30SM300	300,00	30,0	424,4	1			1						25,0		1		1		1	1	17,0
NL	BOHEM	NBK 300	300,00	26,0	576,9	1	2		2				5,0	13,0			1		1		1	1	22,0
NL	FDO	Newecs-25	300,00	26,5	544,0	1	2	1	1	18	NACA 4412/4424		6,0	14,0	25,0	56,0	16		1		1	1	35,0
UK	VAWT	VAWT 100 kW	100,00	16,8	463,0	2	2	4	9		2 NACA 0018		5,0	13,0	30,0	55,0	2	1	1			1	19,5
UK	VAWT	VAWT 130 kW	130,00	24,2	288,9	2	2	4	9		2 NACA 0015		5,0	11,0	30,0	55,0	3	1	1			2	25,0
UK	WIND	WPI	150,00	20,7	454,5	1	3	1	1		2 LS104xx	2930	4,0	14,0		66,0	2	1	1		1	1	21,5
UK	HAWKER	200 kW	200,00	20,8	601,0	1	3	1	1		1		2,7	16,0			4	4			1		
UK	WEGROU	MS-1	250,00	20,2	796,2	1	2	1	1	18	4 NACA 44xx	16000	1,0	17,0	27,0	65,0		1	1		1	3	16,3
UK	WEGROU	MS-2	250,00	25,0	520,0	1	3	1	1	18	1 NASA LS 1	9000	5,0	13,0	25,0	60,0	1	1	1		1	1	25,0
UK	WEGROU	MS-3	300,00	33,0	358,1	1	2	1	1	18	1 NASA LS-1 (mod)	16000	5,0	11,5	25,0	60,0	1	1	1		1	1	30,0

Table 5 : WECs in the rated power range 300.1 - 900 kW

CNT	MANUF.	TYPNAME	RATPOW kW	DIAM m	SPECPOW W/m²	ROTTYP #	NUMBLAD #	ORIENT #	POWCONT *	SPEED #	PROFILE	WEIHEAD kg	VCUTIN m/sec	VRATED m/sec	VCUTOUT m/sec	VSURV m/sec	GENTYP *	TRANS *	BRKTYP *	BRKPOS #	TOWTYP *	TOMAT *	TOWHEI m
B	HMZ	WindMaster 500	500,00	32,5	615,3	1	3	1	50				5,0	13,0			4	4	1				
DK	WENERG	M24 (350kW)	350,00	24,0	789,9	1	3	1	1				4,0	14,0			4	4	1			1	
DK	MICON	M 400kW	400,00	26,0	768,8	1	3	1	1														
DK	DWT	WINDANE 34/400	400,00	34,8	420,6	1	3	1	18	1	NACA 4412-4424	18300	5,0	13,0	25,0	56,0	18	4	1	1	3	1	30,0
DK	WENERG	M24 (400kW)	400,00	24,0	902,7	1	3	1	1				4,0	14,0			4	4	1		3	1	32,6
DK	BONUS	Bonus 450 kW	450,00	35,0	467,8	1	3	1	1	1		21500	5,0	18,0	28,0	67,0	18	3	1		3	1	36,0
DK	NOTANK	NTK 450/37	450,00	37,0	431,4	1	3	1	1												3		
DK	STATSM	NIBE A	630,00	40,0	516,9	1	3	1	1				6,5	14,5	16,5		2				3	2	
DK	STATSM	NIBE B	630,00	40,0	516,9	1	3	1	18				5,5	14,5	18,5		2				3	2	
DK	DWT	WINDANE 40/750	750,00	40,0	596,7	1	3	1	18	1	NACA 4412/4434	70000	3,0	14,5	25,0	70,0	18	4	1	1	3	1	42,0
DK	ELKRAF	750kW	750,00	40,0	609,3	1	3	1	18	1			5,0	15,0	25,0		132	4			1	2	41,0
DK	DENCON	D-750	750,00	26,0	1456,3	1	3	1	1														
FRG	BSW	VR375kW	375,00	36,0	376,1	1	3	1	1	1	NACA 63-200		3,5	13,0	24,0	68,0	4	2	1		8	1	
FRG	TACK	TW 500	500,00	35,0	535,7	1	3	1	1	1			4,0	14,5	20,0	60,0			1		3	1	
FRG	MBB	Monopteros 50	640,00	56,0	265,3	1	1	2	18	4	FX-84Wortmann	58600	4,5	11,0			1		1		3	1	
NL	POLENK	WPS30/4	420,00	30,0	606,6	1	3	1	1				6,0	14,0			4	4			1	1	
NL	NEWINC	AERO34P1500	500,00	34,0	562,2	1	3	1	18		NACA		4,0		25,0		4	4	1		1	1	
NL	NEWINC	AEROTECH34P1500	500,00	34,0	567,7	1	3	1	18								2						
NL	NEWINC	35 SI 500	500,00	35,0	535,7	1	3	1	1	1			5,0	16,0			17		1		3		
NL	HOLEC	WPS-35-550 kW	550,00	35,0	589,3	1	3	1	18	1	NACA 230 family	32000	5,0	16,0	25,0	70,0	6	2	1	1	5	1	37,0
NL	BOHEM	NBK 600	600,00	36,0	601,8	1	2	1	2				5,0	13,0			1		1		1		
UK	HOWDEN	HWP-330	330,00	31,0	446,4	1	3	1	20			22400	5,5	14,0	26,0		5	1	1		1	1	25,0
UK	VAWT	VAWT 850/500kW	500,00	35,0	588,2	2	2	1	1		NACA 0018		7,0		23,0							2	30,0
UK	HOWDEN	HWP-750	750,00	45,0	481,4	1	3	1	20			51000	5,5	14,0	26,0		5	1	1		1	1	35,0
AU	VILLAS	FLODA 600	600,00	36,0	607,7	1	3	1	18	4	FX Wortmann	24000	4,0	14,0	35,0	67,0	17	4	1	2	3	3	42,0
																					3	3	42,0

Table 6 : WECs in the rated power range > 900 kW

CNT	MANUF.	TYPNAME	RATPOW	DIAM	SPECPOW	ROTTYP	NUMBLAD	ORIENT	POWCONT	SPEED	PROFILE	WEIHEAD	VCUTIN	VRATED	VCUTOUT	VSURV	GENTYP	TRANS	BRKTYP	BRKPOS	TOWTYP	TOMMAT	TOWHEI
			kW	m	W/m²	#	#	#	*	#		kg	m/sec	m/sec	m/sec	m/sec	*	*	*	#	*	*	m
DK	ISVEST	Esbjerg 1	2000,00	61,0	682,6	1	3	1	18	1	NACA 4412-43		5,0	15,0	25,0		18	2	1		3	2	57,0
FRG	MAN	WKA 60	1200,00	60,0	433,3	1	3	1	18	4	NACA 44xx	187000	4,9	12,2	24,0	65,0	17	3	1	1	3	2	50,0
FRG	MBB	Aeolus WTS-75	3000,00	75,0	679,1	1	2	1	18		NACA 64-612		0,6		25,0	45,0	2		1	1	1	3	92,0
FRG	MAN	Growian	3000,00	100,4	378,9	1	2	2	18	4	FX-77-W-series	420000	5,4	12,0	24,0	60,0	18	2	1	2	5	1	100,0
I	AERIT	GAMMA 60	1500,00	60,0	541,3	1	2	1	34	4	NACA 230xx		5,0	13,3	27,0	64,0	5	1	1		1	1	61,6
NL	FDO	Newecs-45	1000,00	45,0	628,8	1	2	1	18		LS1-0417/0421	45000	6,0	15,0	20,0	56,0	16	2	1		1	1	60,0
UK	HOWDEN	Richborough	1000,00	55,0	433,9	1	3	1	4	1		80	5,0	13,0	25,0		17	2	1	1	3	1	45,0
UK	VAWT	VAWT 1200KW	1200,00	70,0	318,3	2	2	2	65				8,0	13,0			4		1				
UK	WEGROU	LS-1	3000,00	60,0	1083,2	1	2	1	20	1	NACA 44xx	131000	7,0	17,0	27,0	70,0	1	1	1		1	2	45,0
UK	WEGROU	LS-1	3000,00	60,0	1083,2	1	1	1	4	1	NACA 44xx	131000	7,0	17,0	27,0	70,0	1	1	1			2	45,0
S	STAENE	Naesudden	2000,00	75,0	462,2	1	2		18										1				
S	BOVING	WTS 75-3	3000,00	75,0	693,3	1	2		18										1				
S	STAENE	Maglarp	3000,00	78,0	641,0	1	2		18										1				

5.2   Design Wind Speeds

WECs start operating at a certain wind speeed called v_{cut-in} which is usually between 3 and 5 m/s, depending on the WEC size. Operation at lower windspeeds is not useful because of the relative high losses caused by bearing friction, transmission and generator efficiency.

The power increases with the cube of wind speed, see equ. (1) section 4.1, according to the individual characteristic up to the rated wind speed v_{rated} at which point the generator reaches its rated power. This working point, for which the aerodynamic design of a WEC is optimized, is located on the left before the maximum on the c_p/lambda - characteristic. Above the rated windspeed, the power output of a WEC will remain constant for pitch controlled machines (mostly) by reducing the angle of attack of the rotor blades. Stall controlled WECs show a decreasing power output above v_{rated}, depending on the aerodynamic layout of the fixed rotor blades. The choice of specific rated power, which is the rated power v_{rated} divided by the rotor swept area, depends on the expected wind potential for which the design of a WEC is optimized. High wind potential leads to high v_{rated}. In Fig. 40 the specific rated power is plotted versus v_{rated} for several WECs.

To protect a WEC from destruction at high wind speeds, the safety system stops rotor and generator operation or turns the machine away from the wind by the yaw system at a wind speed called $v_{cut-out}$. There is no power output above $v_{cut\ out}$.
The structural layout of rotor blades and tower depends also on $v_{survival}$. The stationary rotor blades and the tower have to withstand the aerodynamic loads at $v_{survival}$ wind speeds which are usually in the range from 50 to 60 m/s. Thus, the total extractable power from the wind is dependent on the three design wind speeds

$$- v_{cut-in}$$
$$- v_{rated}$$
$$- v_{cut-out}$$

These speeds are determined individually for each WEC type. The available energy potential cannot be extracted by the WEC at wind speeds below v_{cut-in} and above $v_{cut-out}$, and is reduced due to the power limitation at v_{rated}.

The integral of the product of power curve and the probability function over time defines the energy extracted from the wind.

The losses incurred by v_{cut-in}, v_{rated} and $v_{cut-out}$ can be defined in terms of an efficiency parameter η . This parameter η quantifies these losses and is the quantity of total energy in a free wind flow minus the energy loss divided by the total energy of the free wind flow. The energy losses qualitatively represented in Figure 11.

48

ENERGY LOSSES BY POWER LIMITATION

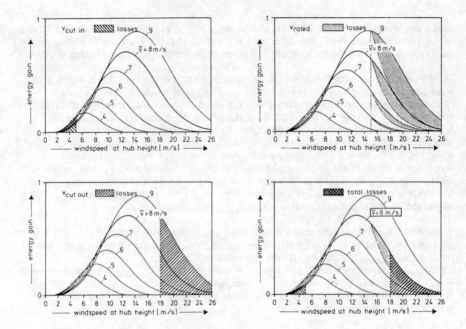

Fig. 11: qualitative presentation of energy losses caused by power limitation
through design wind speeds, parametrized to the mean annual wind speed

The energy losses as a result of the design wind speed parameters of a WEC are as follows :

$\eta_{\text{cut-in}}$ = energy losses caused by $v_{\text{cut in}}$
see Fig.11, upper left

η_{rated} = energy losses caused by v_{rated}
see Fig.11, upper right

$\eta_{\text{cut-out}}$ = energy losses caused by $v_{\text{cut out}}$
see Fig.11, lower left

η_{total} = total losses caused by $v_{\text{cut-in}}$, v_{rated}, $v_{\text{cut-out}}$,
see Fig.11, lower right

Based on a Rayleigh distribution, see equ. (12) section 4.1, and with following substitutions

$$b = \frac{\pi}{4} \cdot \frac{1}{\bar{v}^2}$$

$$x = v$$

$$y = \frac{\sqrt{\pi}}{2} \cdot \frac{v}{\bar{v}}$$

$$d = \frac{\rho}{4} \cdot \pi \cdot T \cdot \frac{1}{\bar{v}^2}$$

the energy losses due to the three design windspeeds v_{cut-in}, v_{rated} and $v_{cut-out}$ can be indicated as

$$e_{cut-in} = e_0 - e_{loss,v_{cut-in}} \tag{16}$$

$$e_{rated} = e_0 - e_{loss,v_{rated}} \tag{17}$$

$$e_{cut-out} = e_0 - e_{loss,v_{cut-out}} \tag{18}$$

$$e_{cut-in} = d \cdot \frac{1}{b^{\frac{5}{2}}} \cdot \int_{v_{cut-in}}^{\infty} y^4 \cdot e^{-y} \, dy \tag{19}$$

$$
\begin{aligned}
e_{rated} \quad &= d \cdot \frac{1}{b^{\frac{5}{2}}} \cdot \int_0^{v_{rated}} y^4 \cdot e^{-y} \, dy \\
&+ d \cdot \frac{1}{b^{\frac{5}{2}}} \cdot v_{rated}^3 \cdot \int_{v_{rated}}^{\infty} y \cdot e^{-y} \, dy
\end{aligned}
\tag{20}
$$

$$e_{cut-out} = d \cdot \frac{1}{b^{\frac{5}{2}}} \cdot \int_0^{v_{cut-out}} y^4 \cdot e^{-y} \, dy \tag{21}$$

The mathematical solution and the quantitative presentaion of

equation   (19)   is shown in   Fig. 12

            (20)   "   Fig. 13

            (21)   "   Fig. 14

for different mean annual windspeeds on the basis of a Rayleigh distribution.

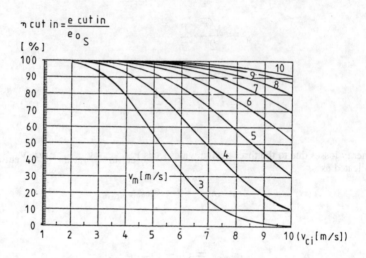

Fig. 12 : energy losses due to v_{cut-in} thresholds for specified MAWS

As shown inFfigure 12, the lower the v_{cut-in}, the less effect the MAWS will have on losses. For example with a v_{cut-in} of 5 m/sec the efficiency is approximately 82% for a MAWS of 4 m/s.
At a mean annual wind speed of 6 m/s, the efficiency is nearly 100%. For higher v_{cut-in}, (eg. 6 m/s) the MAWS has a greater effect on the performance of a WEC.

For economic operation WECs are placed on sites with MAWS of at least 4 m/sec. The cut-in wind speeds are mostly lower than 5 m/sec. From Figure 12 it can be seen that the energy losses due to WECs operating in these conditions are less than 18 % of the total wind energy potential.

Figure 13 shows the energy losses caused by v_{rated}.

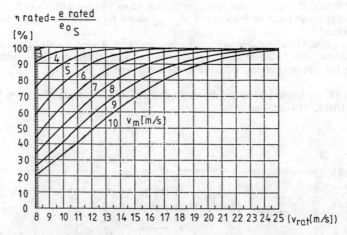

$$\eta \, rated = \frac{e \, rated}{e_o \, S}$$

Fig. 13: Energy losses due to v_{rated} thresholds for specified MAWS

Figure 13 differs from Figure 12 in that it shows that the energy losses decrease with increasing v_{rated}. This is because v_{rated} is located to the right of the maximum on the Rayleigh wind speed distribution curve, see Fig. 8.

Fig. 14 shows the energy losses caused by $v_{cut-out}$.

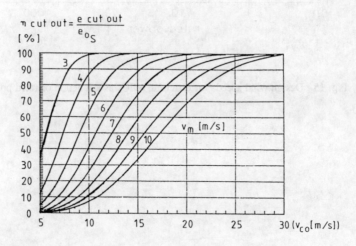

$$\eta \, cut \, out = \frac{e \, cut \, out}{e_o \, S}$$

Fig. 14: Energy losses due to $v_{cut-out}$ thresholds for specified MAWS

As expected, the energy losses in Fig. 14 decrease with increasing $v_{cut-out}$. For most of the newer WECs, the cut-out wind speed approaches 25 m/sec. For MAWS of less than 8 m/sec the energy losses incurred from such a $v_{cut-out}$ threshold are smaller than 3 % of the energy gain potential.

For typical wind turbine sites in Europe with mean annual wind speeds between 4 and 7 m/s one can state that, based on a Rayleigh distribution, $v_{cut-out}$ design wind speeds higher than 20 m/s have an insignificant influence to the annual energy production.

Figure 15, the last in this series, shows the real design wind speeds of existing WECs stored in the EUROWIN database.

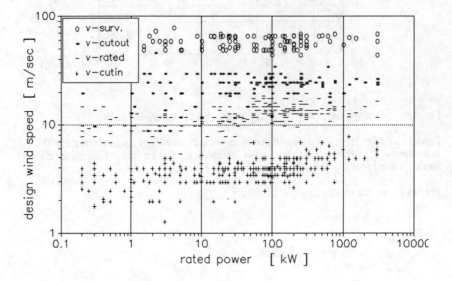

Fig. 15 : Design wind speeds for different WECs as a function of rated power

5.3   Power Curves

The power curve of a WEC is one of its most important characteristics and is the basis for sensitivity analyses such as theoretical calculation of annual energy, economics and more. Power can be defined in many ways ie. power captured by rotor, power into generator, generated power. An unequivocal definition and indication of the meaning of the power curve from the manufacturer side does not yet exist. An objective way to indicate the power curve of a WEC is to plot the electrical power output as a function of wind speed at hub height and this is used in EUROWIN.

In the technical reports one can find measured as well as theoretically calculated power curves. The measurement of power curves entails a large expenditure on measuring equipment and is mostly carried out by institutions such as
- Riso/Denmark
- ECN in Petten/Netherland
- NWTC in Glasgow/UK,
- CRES in Koropi/GR
- DEWI in Wilhelmshaven/Germany
- Germanischer Lloyd in Hamburg /Germany.

To receive certifications by one of these institutions a WEC has to be tested under standard conditions over a certain measuring period.
The theoretical calculation of power curves is done by summarizing the aerodynamic forces and velocities of the rotor blades over the radial sections, it also considers the efficiency of sub-components such as bearings, transmission, coupling and generator. Measured and calculated power curves differ from each other in that calculated power curves mostly follow an ideal curve, determined by the aerodynamics of the rotor and its power control mechanism. Measured power curves indicate the spread of power output under real wind conditions. These deviations are due to the 10-min.-averaging of power output and corresponding windspeed measuements.

Table 7 to 14 which follows show the absolute electric power output in [kW] and the corresponding specific power characteristic in [W/m²] for several machines, as indicated by the manufacturers, as a function of windspeed.

Wind energy converters are classified in the following rated power ranges :

class I	:	0.1	-	10 kW	rated power	(see Table	7	and	12)
class II	:	10.1	-	75 kW	"	(see Table	8	and	13)
class III	:	75.1	-	300 kW	"	(see Table	9	and	14)
class IV	:	300.1	-	900 kW	"	(see Table	10	and	15)
class V	:		>	900 kW	"	(see Table	11	and	16)

The following tables include abbreviations which are used in the database for several WEC types. To aid identification, a complete list of manufacturers and WEC types is given in the Appendic.

Some records do not include complete information this is indicated by "na".

Table 7 : Generated power as a function of wind speed of WECs in the rating class 0.1 - 10 kW as indicated by the manufacturer

WEC-typename	rated power [kW]	rotor diam. [m]	tower height [m]	power [kW] at windspeed [m/sec]																										
				2.5	3.5	4.5	5.5	6.5	7.5	8.5	9.5	10.5	11.5	12.5	13.5	14.5	15.5	16.5	17.5	18.5	19.5	20.5	21.5	22.5	23.5	24.5	25.5	26.5	27.5	28.5	29.5
B 25	0.7	2.5	8.0	..	..	0.05	0.13	0.21	0.36	0.48	0.58	0.63	0.66	0.68	0.69	..	..	:	:	:	:	:	:	:	:	:	:	:	:	:	:
B 30	0.9	3.0	8.0	..	..	0.07	0.17	0.29	0.46	0.64	0.75	0.83	0.87	0.89	0.90	..	..	:	:	:	:	:	:	:	:	:	:	:	:	:	:
BWC 1000	1.2	2.8	15.0	..	..	0.04	0.10	0.20	0.36	0.56	0.80	0.92	1.09	1.19	1.00	0.40	0.41	0.45	..	:	:	:	:	:	:	:	:	:	:	:	:
E 305	1.5	3.4	12.0	..	..	0.50	1.20	2.30	3.40	4.60	6.00	7.10	7.80	8.20	8.30	8.00	7.70	7.40	7.20	7.00	6.90	7.00	6.90	7.10	7.00	7.30	7.60	8.00	8.40	9.00	9.50
LMW 2500	2.5	5.0	na	0.17	0.37	0.80	1.22	1.55	1.80	2.10	2.40	2.55	2.60	2.62	..	..	:	:	:	:	:	:	:	:	:	:	:	:	:	:	:
LMW 3600	3.6	5.0	17.5	..	..	0.20	0.80	1.22	1.65	2.20	2.62	3.20	3.50	3.65	3.73	3.80	..	:	:	:	:	:	:	:	:	:	:	:	:	:
LW 5/5	5.0	5.0	na	..	..	0.60	1.90	3.50	5.30	7.40	9.70	12.30	15.30	18.40	20.00	..	:	:	:	:	:	:	:	:	:	:	:	:	:	:
Flair 8	5.0	5.0	18.0	..	..	0.80	1.70	2.70	3.70	4.40	4.40	4.40	4.40	4.40	4.40	4.40	..	:	:	:	:	:	:	:	:	:	:	:	:
E 710	8.6	7.0	18.0	..	..	0.35	0.85	1.35	2.20	2.95	3.95	4.56	7.85	8.45	8.65	8.50	8.00	7.60	7.40	7.10	6.90	6.60	6.50	6.40	6.40	6.40	6.40	6.40	6.40	6.40
PG 10	10.0	6.3	18.0	..	..	0.40	0.80	1.50	2.50	3.80	5.07	6.01	0.00	13.30	..	:	:	:	:	:	:	:	:	:	:	:	:	:	:	:
BWC EXCEL	10.0	10.0	15.0	..	..	0.20	0.75	1.30	2.30	3.40	4.30	6.00	7.50	9.15	10.00	10.00	8.20	3.15	3.20	..	:	:	:	:	:	:	:	:	:	:
LMW 10/7	10.0	7.0	23.5	..	..	0.20	0.70	1.40	2.20	3.30	4.70	6.00	7.30	9.10	10.00	10.00	8.90	3.30	3.40	..	:	:	:	:	:	:	:	:	:	:
T 103 10/8	10.0	8.0	na	..	..	0.30	0.70	1.20	1.70	2.60	3.70	5.40	7.10	8.80	9.80	10.00	9.65	9.30	..	:	:	:	:	:	:	:	:	:	:	:

na = data not available

-99.00 = data uncomplete

list of complete WEC-type names, see Appendix

Table 8 :   Generated power as a function of wind speed of WECs in the rating class 10.1 - 75 kW
             as indicated by the manufacturer

WEC-typname rated rotor tower power [kW] at windspeed [m/s]

WEC-typname	power [kW]	diam. [m]	height [m]	2.5	3.5	4.5	5.5	6.5	7.5	8.5	9.5	10.5	11.5	12.5	13.5	14.5	15.5	16.5	17.5	18.5	19.5	20.5	21.5	22.5	23.5	24.5	25.5	26.5	27.5	28.5	29.5
Eakel 11 kW	11.0	10.6	na			0.60	1.90	3.50	5.30	7.40	9.70	12.30	15.30	18.40	20.00	-99.00															
Elektromat 12	12.0	6.3	na			0.30	0.65	1.05	1.60	2.40	3.40	4.50	5.50	6.00	7.90	9.00	10.20	11.40	12.00	11.30	10.70	10.40	10.10	9.00							
N 715	13.4	7.0	11.0	0.10	0.51	1.72	2.33	4.45	5.97	7.44	9.51	11.12	12.44	13.35	14.16	14.46	14.76	14.76	14.66	14.46	14.16	13.96	13.75	13.65	13.55	13.65	13.65	13.85	13.96	14.16	
WPS 10	15.0	5.6	na	0.10	0.35	2.05	3.45	5.00	6.15	8.15	11.00	13.70	15.00	-99.00																	
WPS10A15	15.0	9.6	17.0			1.10	2.30	3.90	6.15	9.00	11.70	14.10	15.20	-99.00																	
WPS15SN15	15.0	9.6	na		0.55	1.30	2.45	3.35	6.05	6.05	11.20	13.60	15.10	15.30	-99.00																
MINDANE 10/18	18.5	9.6	18.0		0.80	2.40	4.10	6.00	8.00	10.00	12.00	14.00	15.00	16.50	17.00	17.60	18.20	17.70	16.50	15.20	14.80	14.50	14.10	14.20	14.30	14.50	14.50				
REY 4/18.5	18.5	10.7	18.7	0.20	1.20	2.70	5.50	7.10	9.40	10.50	14.00	17.10	18.50	17.00	16.60	16.50	15.50	19.40	19.00	18.50	18.50	18.50									
Aeolus 11	11.0	11.7	18.0		1.20	2.60	5.00	7.60	10.50	14.00	17.10	18.50	14.00	17.50	18.50	18.50	18.50	18.50	3.50	18.50											
Elektromat 20	20.0	10.0	na	0.70	1.40	2.60	4.20	6.50	9.40	7.90	17.50	19.90	-99.00																		
Twin	20.0	10.6	na	0.50	3.00	6.00	8.64	14.10	18.50	25.71	32.50	48.91	47.80	45.90	43.35	50.00	50.00	49.33	47.70												
Aeroman 12/20	20.0	12.0	18.0		0.30	1.70	6.50	9.30	12.20	16.10	18.20	20.00	20.00	20.00	20.00	20.00	20.00	20.00	20.00	20.00	20.00	20.00	20.00	20.00	20.00	20.00					
E 1220	20.0	12.5	18.5	2.00	4.00	6.50	9.50	13.50	17.00	18.00	19.00	20.00	19.00	18.50	16.50	15.00	14.50	14.00	14.00	13.50	13.50	13.00	13.00	13.00	13.00	13.25	13.30				
N 22	22.0	9.8	18.0	2.00	4.00	7.00	9.00	13.00	18.00	20.00	21.00	21.00	21.00	21.00	21.00	21.00	21.00	21.00	21.00	21.00	28.00	28.00	28.00	28.00	28.00	28.00	28.00	28.00	28.00	20.00	
Monopterus 20	25.0	12.5	15.0		2.20	5.00	7.00	11.40	15.00	17.60	20.44	22.20	23.60	24.30	25.10	25.20	25.20														
LW 25 kW	25.0	10.6	na	1.29	2.00	4.65	7.14	9.62	12.44	15.70	19.40	23.30	21.50	22.10	22.71	23.40															
Cauperduin	30.0	10.6	na	0.40	2.30	5.35	8.35	11.44	14.70	17.00	18.00	19.50	20.50	21.50	22.71	23.40	23.40														
12/30	30.0	12.0	14.4			4.00	7.00	11.00	14.00	18.00	22.00	26.00	29.50	29.50	29.50	29.50	27.00	27.00	27.00	27.00	27.00	26.00	25.00	25.00	25.00	25.00	25.00	25.00			
Aeroman 12.5/30	30.0	12.5	18.0			3.00	7.00	11.00	15.00	20.00	27.00	30.00	30.00	31.00	32.00	32.25	32.50	32.00	31.00	30.50	30.40	30.30	30.20	30.10							
Darrieus DZ-12	30.1	12.0	18.0			2.17	7.50	12.00	15.00	22.00	27.00	30.00	30.00	30.00	30.00	30.40	30.40	30.40	30.40	30.40											
F.M.W. 30	30.1	12.5	15.0	1.70	3.00	5.00	7.60	11.00	15.30	20.20	26.60	30.40	30.40	30.40	30.40																
N 12/30	30.0	12.5	18.0	2.50	5.00	8.00	11.50	17.50	24.60	30.50	32.00	30.00	30.00	27.00	25.50	23.00	21.50	18.00	17.00	16.50	16.00	16.50	16.50	16.50	16.50	17.00	17.50				
Kano-Rotor 30kW	30.4	12.5	15.5	1.07	2.30	5.07	6.12	9.14	14.05	18.67	22.36	26.04	30.31	32.14	33.58	33.67	32.76	32.45	32.60	33.55	35.51										
HSW-30	33.1	12.5	14.5		2.00	5.00	10.00	15.00	20.00	25.05	30.00	32.50	32.50	32.50	32.50	32.50	32.50	32.50	32.50	32.51											
Aeroman14.8/33n	33.1	14.8	15.0	2.13	4.26	8.51	14.04	20.00	26.53	29.51	31.51	32.77	32.77	32.77	32.77	32.77	32.77	32.77	32.77	32.77											
WPS11A40	41.0	11.5	17.0	1.63	3.10	5.28	8.25	12.10	16.30	20.50	24.60	29.00	23.50	36.75	40.00	-93.00															
N 1245	45.0	11.5	18.5	2.50	5.00	9.00	13.50	18.00	31.50	37.50	43.00	45.00	45.00	39.00	36.00	32.00	28.50	26.00	23.50	21.50	21.50	17.50	16.00	14.50	13.50	13.00	13.00	12.50	11.50	11.00	
TW 45	45.1	12.5	na	0.50	2.50	6.00	10.00	14.00	20.00	27.00	35.00	44.00	49.00	50.50	52.00	51.50	51.00	51.50	52.30												
BOM 55/11	55.0	11.0	na	2.00	3.50	7.50	15.00	22.50	37.50	35.00	50.00	57.00	58.00	60.00	60.00	60.00	60.00	60.00	60.00												
VEST 55/15.3	55.0	15.3	na	3.00	7.00	12.00	18.10	26.00	42.00	48.00	53.00	55.00	55.00	55.00	58.00	58.00	58.00	58.00													
(55/16)	55.0	16.0	20.0		1.00	5.50	10.50	17.50	26.00	43.00	46.00	56.50	60.50	60.50	60.50	61.50	61.50														
REY 55/15.6	55.1	15.6	na	2.00	5.00	10.00	15.00	22.00	35.00	45.00	52.00	55.00	55.00	55.00	55.00	50.00	47.00	46.00	44.00												
WPS16A60	60.0	15.6	na	3.33	6.16	11.10	17.20	24.40	32.50	40.00	47.20	54.70	59.10	-99.00																	
TW 60	60.0	16.3	23.5	2.00	7.50	12.10	20.00	28.00	32.50	40.00	45.50	50.00	54.00	57.00	59.00	60.00	65.00	66.00	65.00												
N 66/13-US	65.0	16.3	na	2.00	4.00	10.00	18.00	25.00	38.00	47.00	54.00	58.00	62.00	64.50	65.00	67.00	67.00	67.00	67.00												
WR 65/11	65.0	17.0	na	1.00	2.00	3.50	5.50	9.50	12.00	16.00	22.50	30.00	40.00	45.00	53.00	57.50	65.00	65.00	65.00												
15-15	75.0	15.2	22.5	3.00	10.00	18.00	23.00	38.00	47.00	56.00	60.00	64.00	68.00	75.00	81.00	88.00	91.00	93.00	96.50	97.00	97.50	98.00									
VEST 75/15	75.1	15.0	na	3.00	8.60	13.20	21.20	29.60	41.00	56.00	66.90	67.00	74.30	73.50	74.30	75.20	75.75	75.90	76.00												
NTK 75/16	75.0	16.0	na	3.00	8.00	13.00	21.20	25.80	41.00	56.00	66.90	67.00	74.00	81.00	83.00	91.00	93.00	93.00	93.00	83.00	83.00	83.00	83.00	93.00	93.00	93.00	93.00				

na   = data not available
-99.00   = data uncomplete

list of complete WEC-type names, see Appendix

Table 9 : Generated power as a function of wind speed of WECs in the rating class 75.1 - 300 kW
as indicated by the manufacturer

WEC-typname	rated power [kW]	rotor diam.[m]	tower height [m]	power [kW] at windspeed [m/s] 2.5	3.5	4.5	5.5	6.5	7.5	8.5	9.5	10.5	11.5	12.5	13.5	14.5	15.5	16.5	17.5	18.5	19.5	20.5	21.5	22.5	23.5	24.5	25.5	26.5	27.5	28.5	29.5
ENERCON-17/80	80.0	17.2	21.0	0.3	3.0	5.8	9.8	16.9	25.8	37.3	48.9	62.7	72.9	78.7	80.0	80.0	80.0	80.0	80.0	80.0	80.0	80.0	80.0	80.0	80.0	80.0	:	:	:	:	:
Svedana 81/50Hz	80.0	17.0	24.0	3.0	7.0	10.0	14.0	17.0	22.0	38.0	40.3	55.0	62.0	66.0	69.0	73.0	77.0	78.0	78.0	77.0	75.0	73.0	72.0	71.0	71.0	73.0	75.0	76.0	77.0	:
UK T-1700	80.0	17.0	na	2.0	7.0	12.0	20.0	28.0	35.0	40.0	47.0	57.0	67.0	76.0	78.0	77.5	76.0	73.0	73.0	71.0	72.0	73.0	75.0	77.0	78.0	80.0	81.0	77.0	:	:
WINDANE 17/80	80.0	17.0	22.0	1.5	4.5	8.0	13.0	21.5	30.0	40.0	47.0	57.0	67.0	77.5	80.5	80.5	78.0	75.0	75.0	71.5	68.5	68.5	70.0	72.0	74.0	76.5	79.0	82.0	:	:
MR80/18	80.0	18.0	na	0.5	1.5	3.0	5.5	8.0	11.0	15.4	21.5	30.0	36.0	43.0	63.5	74.2	78.0	80.0	80.0	80.0	80.0	:	:	:	:	:	:	:	:	:
15-90	90.0	15.0	22.5	1.3	3.0	5.0	9.0	14.3	20.0	28.5	38.5	46.5	55.0	62.5	65.5	71.8	75.5	78.0	78.0	78.0	:	78.0	:	:	:	:	:	:	:	:
NOAH90kW	90.0	17.0	na	0.5	3.4	6.0	10.0	14.8	20.0	27.0	33.0	40.0	46.0	44.0	48.0	58.5	52.0	51.5	51.0	51.5	52.3	:	:	:	:	:	:	:	:	:
AWP 80/18	90.0	18.0	22.5		4.0	10.5	22.0	34.0	45.0	55.0	55.4	64.0	71.8	80.5	87.1	91.0	95.3	95.7	94.0	93.3	90.0	86.2	86.2	87.1	:	:	:	:	:	:
V15-90	90.0	18.0	23.0		4.0	10.5	22.0	34.0	45.0	55.0	55.4	64.0	72.0	78.0	82.0	86.0	86.0	86.0	86.0	86.0	86.0	86.0	86.5	85.0	:	:	:	:	:	:
Svedana100/50Hz	95.0	17.0	24.0	1.0	4.0	8.0	13.0	18.0	26.0	38.0	54.0	63.0	74.0	81.0	89.0	92.5	95.0	94.0	92.0	91.0	88.0	88.0	88.0	91.0	91.0	92.0	93.0	93.0	93.0	94.0
WINDANE 19/95	95.0	19.0	22.0		2.0	5.0	9.0	14.0	20.6	31.3	41.7	53.0	71.5	90.0	94.0	95.0	95.0	94.0	92.0	89.0	88.0	83.5	84.0	86.0	87.5	88.0	92.0	92.0	94.0	:
WPS11A100	100.0	18.0	17.0		3.3	7.3	12.6	20.6	31.3	41.7	53.0	64.7	76.7	86.7	94.0	92.5	95.0	91.0	92.0	91.0	90.0	83.0	83.5	84.0	86.0	87.5	:	:	:	:
WPS15M100	100.0	18.4	na	1.4	6.4	13.6	21.0	32.0	43.0	56.3	71.7	86.7	97.5	99.0	97.5	-99.0	100.0	:	:	:	:	:	:	:	:	:	:	:	:	:
BON 100/19.4	100.0	19.4	23.5	4.0	9.0	16.2	30.7	43.7	58.1	73.8	87.0	95.5	102.4	107.1	112.5	102.5	112.5	112.0	103.7	110.2	112.5	112.5	112.5	112.5	112.5	112.5	112.5	112.5	:	:
BON 100/20	100.0	20.0	30.0	3.6	9.3	17.0	27.3	43.8	58.1	73.4	86.0	94.6	101.7	103.3	111.4	111.0	112.0	112.5	112.0	110.0	112.5	112.5	112.5	112.5	112.5	112.5	110.0	:	:	:
W 100 (V.2)	100.0	19.0	22.0	6.0	11.0	20.0	35.0	53.0	68.0	86.0	88.0	100.0	100.0	100.0	110.0	110.0	110.0	110.0	110.0	110.0	105.0	105.0	105.0	:	:	:	:	:	:	:
20-100	100.0	18.0	30.0				9.5	20.0	24.0	46.0	59.0	70.0	83.0	88.0	94.5	99.0	104.5	100.0	100.0	100.0	100.0	100.0	100.0	100.0	100.0	:	:	:	:	:
V 20-100kW	100.0	20.0	23.4		1.0	4.5	11.0	21.0	35.0	47.5	62.5	71.5	77.5	82.0	84.3	93.0	104.5	107.0	108.5	109.0	108.0	107.5	105.0	101.5	103.0	:	:	:	:	:
WPS 20	100.0	20.0	na	4.0		17.0	30.0	47.0	67.0	83.0	95.0	100.0	100.0	100.0	105.0	113.0	121.0	122.0	115.0	:	:	:	:	:	:	:	:	:	:	:
ADLER 25/100kW	100.0	25.0	22.0		2.0	5.0	12.0	24.0	40.0	58.0	74.0	87.0	95.0	100.0	100.0	100.0	95.0	-95.0	:	:	:	:	:	:	:	:	:	:	:	:
M 100	100.0	19.3	na	1.4		5.0	12.0	24.0	40.0	58.0	74.0	87.0	95.0	104.0	107.5	111.0	112.0	112.0	112.0	112.0	112.0	123.0	123.0	:	:	:	:	:	:	:
W-150	110.0	19.0	23.0	1.1	5.1	15.1	31.5	44.3	61.0	61.0	85.0	100.4	112.5	116.2	120.4	121.6	120.0	123.7	123.0	123.0	123.0	123.0	:	:	:	:	:	:	:	:
NTK - 130F	130.0	20.5	24.0	4.3	9.2	20.2	37.7	54.0	69.3	83.0	95.0	100.8	121.5	130.7	131.0	141.1	141.1	141.0	139.3	139.7	138.5	135.0	134.4	133.7	131.3	130.1	130.3	130.0	130.0	130.0
NTK 130/20.4	130.0	20.4	na	1.1	6.1	21.0	37.0	50.0	65.0	80.0	95.0	100.0	112.0	116.0	123.0	126.0	128.0	130.0	130.0	131.0	132.0	132.0	:	:	:	:	:	:	:	:
D 140	140.0	19.6	na	2.5	5.3	15.5	29.3	45.0	63.0	83.0	100.3	109.0	121.0	130.0	135.0	140.0	141.0	142.2	142.5	142.5	141.5	140.5	:	:	:	:	:	:	:	:
DANWIN 19/140	140.0	19.0	22.5	5.0		9.0	16.0	27.0	41.0	65.0	83.0	95.0	100.0	120.0	130.0	135.0	140.0	139.0	139.0	139.0	139.0	139.0	139.0	130.0	:	:	:	:	:	:
20/150	150.0	20.5	30.0	2.0	13.0	26.0	41.0	60.0	82.0	68.0	114.0	134.0	152.0	156.0	147.0	154.0	152.0	150.0	143.0	143.0	142.0	148.0	148.0	148.0	148.0	148.0	:	:	:	:
TW 150	150.0	20.5	na			9.5	20.0	32.0	41.0	60.0	77.0	100.0	125.0	150.0	152.5	152.5	152.5	156.0	158.0	159.0	159.0	159.0	159.0	:	:	:	:	:	:	:
D 150	150.0	22.2	23.5	3.0	8.0	17.0	31.5	50.0	73.0	100.0	122.0	137.0	144.0	145.0	146.0	145.0	144.0	142.4	138.5	121.5	116.5	112.0	109.0	105.0	105.0	102.5	:	:	:	:
V-S 150 (V.2)	150.0	21.0	25.5	0.4		13.7	20.5	30.0	44.5	58.1	77.1	101.1	122.4	132.7	142.7	148.7	148.1	144.0	142.4	135.5	126.5	117.5	106.3	107.0	107.0	107.0	:	:	:	:
Windmaster 150	150.0	21.0	21.0			12.0	18.0	26.0	34.0	46.0	70.0	86.0	85.0	117.0	143.0	150.0	150.0	150.0	150.0	150.0	150.0	150.0	150.0	:	:	:	:	:	:	:
BONUS 150 kW	150.0	23.0	29.5	6.6	17.2	23.7	45.3	65.7	95.1	118.0	136.5	149.2	153.3	150.0	143.0	135.2	127.0	122.1	118.4	118.0	117.2	117.2	117.2	117.2	117.2	:	:	:	:	:
V-S 150 (V.3)	150.0	23.0	30.5	1.4	6.6	13.7	20.5	34.0	55.0	77.0	101.5	122.4	128.7	140.2	148.0	142.4	135.5	126.5	127.0	111.0	107.0	106.0	107.0	:	:	:	:	:	:	:
NTK 150 XLR	150.0	24.6	30.0	4.0	12.5	35.0	54.0	75.0	100.0	125.0	134.0	152.0	158.0	150.0	150.0	150.0	150.0	150.0	150.0	150.0	150.0	150.0	150.0	150.0	:	:	:	:	:	:
NORDEX 150 kW	150.0	27.0	30.0	7.0	18.5	39.0	68.5	103.0	129.5	145.0	154.5	160.0	162.0	161.5	159.5	156.5	154.0	152.5	152.5	159.5	165.5	174.5	186.0	:	:	:	:	:	:	:
BOUMA 160/20	160.0	20.0	na		7.5	16.5	30.0	45.0	54.5	74.5	90.0	113.5	128.0	151.0	161.3	164.0	163.0	161.0	161.0	161.0	160.0	160.0	95.0	93.0	100.0	:	:	:	:	:
LN 11	160.0	23.0	na	5.0	11.0	30.0	47.0	67.1	97.0	109.0	132.0	151.0	163.0	164.0	165.0	:	:	:	:	:	:	:	:	:	:	:	:	:	:	:
ADLER 25/165kW	165.0	25.0	22.0	4.0	9.0	17.0	30.0	47.0	67.1	91.0	105.7	132.1	154.6	170.3	175.3	160.8	160.4	150.4	140.2	132.6	127.6	127.0	:	:	:	:	:	:	:	:
V-S 110	180.0	22.2	25.0		12.0	24.0	40.0	56.1	77.0	101.0	122.4	142.0	148.0	141.0	141.1	161.1	161.1	225.0	221.0	225.0	225.0	200.0	200.0	:	:	:	:	:	:	:
Windmaster200kW	200.0	21.0	22.5	3.0	15.0	25.0	40.0	60.0	86.0	105.0	133.0	153.0	177.0	195.0	200.0	200.0	200.0	200.0	200.0	200.0	200.0	200.0	200.0	:	:	:	:	:	:	:
200kW/415m^2	200.0	23.0	23.6	3.0	15.0	25.0	40.0	57.0	77.0	107.0	125.0	166.0	191.0	199.0	200.0	200.0	200.0	200.0	200.0	200.0	200.0	200.0	200.0	:	:	:	:	:	:	:

Table 9 : Generated power as a function of wind speed of WECs in the rating class 75.1 - 300 kW   (contd.)
as indicated by the manufacturer

WEC-typname	ratedrotor power [kW]	rotordiam. [m]	tower height [m]	power at windspeed [m/s] 2.5	3.5	4.5	5.5	6.5	7.5	8.5	9.5	10.5	11.5	12.5	13.5	14.5	15.5	16.5	17.5	18.5	19.5	20.5	21.5	22.5	23.5	24.5	25.5	26.5	27.5	28.5	29.5
DANWIN 24	200.0	24.0	23.0	:	:	5.4	16.3	32.4	52.3	76.8	103.0	130.4	155.9	177.2	193.8	200.7	198.3	191.2	178.5	170.2	155.0	151.8	149.4	152.7	158.5	163.9	:	:	:	:	:	:
DANWIN 24 (US)	200.1	24.0	23.0	:	:	5.4	16.3	32.4	52.3	76.1	103.0	130.4	155.9	177.2	193.8	200.7	198.3	191.2	178.5	170.2	156.0	151.3	149.4	152.7	158.5	163.3	:	:	:	:	:	:
V25-200	200.0	25.0	28.7	:	:	7.3	17.5	32.5	52.0	73.0	96.0	121.0	140.5	171.0	192.5	200.0	200.0	200.0	200.0	200.0	200.0	200.0	200.0	200.0	:	:	:	:	:	:	:
V-S 225	225.0	24.0	28.0	:	:	:	:	24.4	42.0	66.3	94.3	126.3	158.7	186.6	207.2	214.3	221.4	217.3	203.6	200.5	188.0	174.0	165.2	158.5	156.6	:	:	:	:	:	:
NORDEX 225/250	225.0	26.0	30.0	:	:	7.0	15.5	31.5	54.5	86.5	125.0	164.5	191.0	219.5	230.5	234.5	235.0	235.0	235.0	235.0	235.0	235.0	235.0	235.0	235.0	:	:	:	:	:
AIT 02 - Medit	225.0	32.0	25.0	:	:	:	35.0	70.0	110.0	155.0	205.0	225.0	225.0	225.0	225.0	225.0	225.0	225.0	225.0	225.0	225.0	225.0	:	:	:	:	:	:	:	:
TW 250	250.0	23.0	28.5	4.0	10.0	21.0	35.0	50.0	72.0	96.0	122.0	154.0	183.0	210.0	230.0	248.0	252.0	232.0	:	:	:	:	:	:	:	:	:	:	:	:
24.5m/250kW	250.0	24.5	25.0	:	13.0	33.0	55.0	80.0	107.0	138.0	165.0	191.0	214.0	235.0	245.0	254.0	256.0	256.0	256.0	256.0	256.0	256.0	256.0	256.0	256.0	256.0	:	:	:
M 450-250kW	250.0	24.0	21.7	5.0	14.0	28.0	53.0	80.0	113.0	142.0	171.0	201.0	230.0	250.0	256.0	255.0	245.0	240.0	227.0	215.0	:	:	:	:	:	:	:	:	:
HSW-250	250.0	25.0	21.3	28.0	50.0	65.0	95.0	130.0	175.0	205.0	225.0	242.0	258.0	265.0	265.0	261.0	256.0	230.0	291.0	292.0	291.0	:	:	:	:	:	:	:	:
MS-2	250.0	25.0	25.0	:	6.0	25.0	46.0	76.0	114.0	152.0	192.0	223.0	250.0	250.0	250.0	250.0	250.0	250.0	250.0	250.0	250.0	250.0	:	:	:	:	:	:
Windmaster 250	250.0	25.0	30.0	:	11.0	28.0	53.0	80.0	120.0	150.0	195.0	223.0	241.0	250.0	250.0	250.0	250.0	250.0	250.0	250.0	:	:	:	:	:	:	:	:
M 530-250kW	250.0	26.0	28.7	:	12.0	33.6	47.1	84.3	135.2	172.2	205.0	231.1	246.2	250.0	243.0	232.8	227.0	227.0	227.0	227.0	227.0	:	:	:	:	:	:	:
V-S 270	270.0	26.0	24.0	4.1	17.4	33.7	54.1	89.3	126.5	163.3	203.1	235.1	250.0	255.1	257.1	257.1	251.0	242.3	231.6	217.3	202.0	193.3	189.6	:	:	:	:	:	:
Windmaster 300	300.0	25.0	21.8	3.2	22.4	48.0	80.0	113.6	158.4	200.0	244.3	278.4	292.0	296.0	300.0	300.0	300.0	300.0	300.0	300.0	300.0	300.0	:	:	:	:	:	:
NEWECS-25	300.0	26.5	22.0	:	4.0	20.0	49.0	90.0	132.0	181.0	254.0	300.0	300.0	300.0	300.0	300.0	:	:	:	:	:	:	:	:	:	:	:	:
NTK 300	300.0	27.5	30.5	:	6.0	25.0	53.0	91.0	132.0	176.0	225.0	270.0	310.0	340.0	356.0	355.0	345.0	330.0	330.0	330.0	330.0	330.0	:	:	:	:	:	:
WPS05M300	300.0	30.0	11.0	7.0	22.0	38.0	59.0	86.0	121.0	165.0	220.0	276.0	302.0	-99.0	:	:	:	:	:	:	:	:	:	:	:	:	:	:
WINDANE 31/300	300.0	30.8	30.8	:	10.0	30.0	55.0	90.0	130.0	185.0	245.0	290.0	300.0	300.0	300.0	300.0	300.0	300.0	300.0	300.0	300.0	300.0	:	:	:	:	:	:
EMERCON 32/300	300.0	32.0	34.5	8.6	18.4	34.3	55.2	84.1	122.2	169.1	225.4	300.0	300.0	300.0	300.0	300.0	300.0	300.0	300.0	300.0	300.0	300.0	:	:	:	:	:	:
MS-3	300.0	33.0	25.0	:	12.0	35.0	65.0	105.0	155.0	225.0	300.0	300.0	300.0	300.0	300.0	300.0	300.0	300.0	:	:	:	:	:	:	:	:	:	:

na   = data not available

-99.00 = data incomplete

For list of complete WEC-type names, see Appendix

Table 10 : Generated power as a function of wind speed of WECs in the rating class 300.1 - 900 kW as indicated by the manufacturer

WEC-typname	rated power [kW]	rotor diam. [m]	tower height [m]	power [kW] at windspeed [m/s] 4.5	5.5	6.5	7.5	8.5	9.5	10.5	11.5	12.5	13.5	14.5	15.5	16.5	17.5	18.5	19.5	20.5	21.5	22.5	23.5	24.5	25.5	26.5	27.5	28.5	29.5
WINDANE34/400	400	34.1	30.0	10.0	40.0	90.0	130.0	188.0	250.0	310.0	370.0	400.0	400.0	400.0	400.0	400.0	400.0	400.0	400.0	400.0	400.0	400.0	400.0	400.0	400.0	400.0	...	...	...
BONUS 450 kW	450	35.0	32.6	14.3	31.5	60.7	97.2	140.5	179.2	235.7	280.4	329.5	378.6	420.5	453.8	479.8	492.8	500.0	495.7	482.7	462.5	439.3	427.8	420.5	419.1	419.1	419.1	...	...
FLORA 600	600	36.0	42.0	17.0	39.0	73.0	125.5	189.0	259.0	338.5	412.5	476.0	534.0	585.0	595.0	600.0	600.0	600.0	600.0	600.0	600.0	600.0	600.0	600.0	600.0	600.0	600.0	600.0
NIBE A	630	40.0	44.5	...	...	10.0	55.0	115.0	185.0	262.5	350.0	444.4	529.5	600.0	629.5	643.2	...	...	...	...	...	...	...	...	...	...	...	...	...
NIBE B	630	40.0	45.0	...	15.0	62.5	115.0	185.0	260.0	350.0	450.0	550.0	595.0	600.0	600.0	600.0	600.0	...	...	...	...	...	...	...	...	...	...	...
WINDANE 40/750	750	40.0	42.0	...	15.0	55.0	100.0	175.0	235.0	330.0	420.0	530.0	620.0	685.0	750.0	750.0	750.0	750.0	750.0	750.0	750.0	750.0	750.0	...	...	...	...	...

Table 11 : Generated power as a function of wind speed of WECs in the rating class > 900 kW as indicated by the manufacturer

WEC-typname	rated power [kW]	rotor diam. [m]	tower height [m]	power [kW] at windspeed [m/s] 4.5	5.5	6.5	7.5	8.5	9.5	10.5	11.5	12.5	13.5	14.5	15.5	16.5	17.5	18.5	19.5	20.5	21.5	22.5	23.5	24.5	25.5	26.5	27.5	28.5	29.5
WKA 60	1200	60.0	50.0	...	60.0	150.0	260.0	390.0	590.0	800.0	1040.0	1200.0	1200.0	1200.0	1200.0	1200.0	1200.0	1200.0	1200.0	1200.0	1200.0	1200.0	1200.0	1200.0	...	...	...	...	...
GAMMA 60	1500	60.0	61.6	...	120.0	200.0	250.0	375.0	500.0	750.0	1000.0	1250.0	1500.0	1500.0	1500.0	1500.0	1500.0	1500.0	1500.0	1500.0	1500.0	1500.0	1500.0	1500.0	1500.0	...	...	...	...
ESBJERG 1	2000	61.0	57.0	...	...	110.0	230.0	400.0	580.0	750.0	990.0	1210.0	1440.0	1620.0	1800.0	1930.0	2000.0	2000.0	2000.0	1800.0	1560.0	1300.0	1050.0	...	...	...	...	...	...
LS-1	3000	60.0	45.0	...	...	...	190.0	420.0	670.0	890.0	1170.0	1450.0	1710.0	2100.0	2320.0	2460.0	2820.0	3000.0	3000.0	3000.0	3000.0	3000.0	3000.0	3000.0	...	...	...	...	...
GROWIAN	3000	100.4	100.0	...	75.0	375.0	750.0	1125.0	1575.0	2137.5	2700.0	3000.0	3000.0	3000.0	3000.0	3000.0	3000.0	3000.0	3000.0	3000.0	3000.0	3000.0	3000.0	3000.0	...	...	...	...	...

na   = data not available

-99.00 = data incomplete

for list of complete WEC-type names, see Appendix

Table 12: Specific power [W/m²] as a function of windspeed of small-scale WECs in the rating class 0.1 to 10 kW
Data obtained from manufacturer's power curve

WEC-typname	rated power [kW]	rotor diam. [m]	tower height [m]	2.5	3.5	4.5	5.5	6.5	7.5	8.5	9.5	10.5	11.5	12.5	13.5	14.5	15.5	16.5	17.5	18.5	19.5	20.5	21.5	22.5	23.5	24.5	25.5	26.5	27.5	28.5	29.5
B 25	0.7	2.5	8.00	:	:	10.4	27.0	43.7	74.9	99.8	120.6	131.0	137.3	141.4	143.5	:	:	:	:	:	:	:	:	:	:	:	:	:	:	:	:
B 30	0.9	3.0	8.0	:	:	10.1	24.6	41.9	66.4	92.4	108.3	119.9	125.7	128.5	13..	:	:	:	:	:	:	:	:	:	:	:	:	:	:	:	:
BMC 1000	1.2	2.8	15.0	:	:	6.6	16.6	33.2	59.7	92.8	132.6	152.5	180.7	197.3	197.3	165.8	66.3	68.0	74.6	:	:	:	:	:	:	:	:	:	:	:	:
E 305	1.5	3.4	12.0	:	:	56.2	134.9	258.6	382.3	517.2	674.7	798.4	877.1	922.1	933.3	899.6	865.8	832.1	809.6	787.1	775.9	775.9	787.1	798.4	820.9	854.6	899.6	944.5	1012.0	1068.2	1147.0
LMW 2500	2.5	5.0	na	8.8	19.2	41.6	63.4	80.6	93.6	109.2	124.8	132.6	135.2	136.2	:	:	:	:	:	:	:	:	:	:	:	:	:	:	:	:	:
LMW 3600	3.6	5.0	17.5	:	:	10.4	41.6	63.4	85.8	114.4	136.2	166.4	182.0	189.8	194.1	197.6	:	:	:	:	:	:	:	:	:	:	:	:	:	:	:
LW 5/5	5.0	5.0	na	:	:	31.2	98.8	182.0	275.6	394.8	504.4	639.5	795.5	956.7	1039.9	:	:	:	:	:	:	:	:	:	:	:	:	:	:	:
Flair 8	5.0	8.0	18.0	:	:	16.2	34.5	54.8	75.2	89.4	89.4	89.4	89.4	89.4	89.4	89.4	:	:	:	:	:	:	:	:	:	:	:	:	:	:	:
E 710	8.6	7.0	18.0	9.3	22.6	35.8	58.4	78.3	104.8	118.0	171.1	208.2	224.2	229.5	225.5	212.2	201.6	196.3	188.3	183.0	180.4	175.1	172.4	169.8	169.8	169.8	169.8	169.8	169.8	169.8	169.8
PG 10	10.0	6.3	18.0	13.1	26.2	49.1	81.9	124.5	180.1	248.9	327.5	435.6	:	:	:	:	:	:	:	:	:	:	:	:	:	:	:	:	:	:	:
BMC EXCEL	10.0	7.0	15.0	:	5.3	19.9	34.5	61.0	90.2	114.1	159.2	199.0	242.7	265.3	265.3	265.3	217.5	83.6	84.9	:	:	:	:	:	:	:	:	:	:	:	:
LMW 10/7	10.0	7.0	23.5	:	5.3	18.6	37.1	58.4	87.5	124.7	159.2	193.7	241.4	265.3	265.3	236.1	87.5	90.2	:	:	:	:	:	:	:	:	:	:	:	:	:
T 103 10/8	10.0	8.0	na	:	6.1	14.2	24.4	34.5	52.8	75.2	109.7	144.2	178.7	199.0	203.1	196.0	188.9	:	:	:	:	:	:	:	:	:	:	:	:	:	:

na = data not available

For list of complete WEC-typnames, see Appendix

Table 13: Specific power [W/m²] as a function of windspeed of small-scale WECs in the rating class 10.1 to 75 kW
Data obtained from manufacturer's power curve

WEC-typename	rated power [kW]	rotor dia. [m]	tower height [m]	2.5	3.5	4.5	5.5	6.5	7.5	8.5	9.5	10.5	11.5	12.5	13.5	14.5	15.5	16.5	17.5	18.5	19.5	20.5	21.5	22.5	23.5	24.5	25.5	26.5	27.5	28.5	29.5
Enkel 11kW	11.0	10.6	na			6.9	22.0	40.5	61.3	85.6	112.2	142.3	177.0	212.9	231.4																
Elektromat 12	12.0	6.3	na	9.8	21.3	34.4	52.4	78.6	111.3	147.4	180.1	222.7	258.7	294.8	334.1	373.4	393.0	370.1	350.4	340.6	330.8	321.0									
N 715	13.4	7.0	18.0				13.5	45.6	77.7	118.0	158.4	199.4	252.3	295.0	330.0	354.1	375.6	383.6	391.6	388.9	383.6	375.6	370.3	364.8	362.1	359.5	362.1	364.8	367.4	370.3	375.6
WPS 10	15.0	9.6	na			15.5	32.4	55.0	86.7	126.9	165.0	198.9	214.4																		
WPS10A15	15.0	9.6	17.0	1.4	13.4	28.9	48.7	70.5	96.6	124.8	155.2	193.2	211.6																		
WPS10SA15	15.0	9.6	na	7.8	18.3	34.6	55.7	85.3	122.0	158.0	191.8	213.0	215.8																		
WINDANE 10/18	18.5	9.6	18.0				11.3	33.9	57.8	84.6	112.8	141.1	169.3	197.5	218.6	242.6	249.1	256.7	248.2	239.8	232.7	222.8	214.4	208.8	204.5	200.3	198.9	200.3	201.7	204.5	208.8
REY 4/18.5	18.5	10.7	18.7	2.3	13.6	30.6	62.5	88.6	106.7	122.6	138.5	151.0	164.6	178.2	188.5	202.1	215.7	220.3	221.4	212.3	210.0	208.9									
Aeolus 11	18.0	11.7	10.0		11.4	26.6	47.5	72.2	99.7	132.9	169.0	195.7	175.7	175.7	175.7	175.7	175.7	175.7	175.7	175.7	175.7	175.7									
Elektromat 20	20.0	10.0	na	9.1	18.2	33.8	54.6	84.5	122.2	102.7	227.5	258.7																			
Twin	20.0	10.6	na	10.4	34.7	69.4	111.1	163.1	225.6	297.3	380.6	470.9	553.0	577.3	577.9	578.5	570.4	551.8													
Aeroman 12/20	20.0	12.0	10.0		8.1	33.4	58.7	84.0	110.1	145.3	173.3	180.5	180.5	180.5	180.5	180.5	180.5	180.5	180.5	180.5	180.5	180.5	180.5	180.5	180.5	180.5	180.5	180.5	180.5	180.5	180.5
E 1220	20.0	12.5	18.5		16.6	33.3	54.1	74.9	112.3	141.4	149.8	158.1	166.4	158.4	149.8	145.6	137.3	133.1	124.8	120.6	116.5	112.3	108.2	108.2	110.2	110.7					
N 22	22.0	9.8	18.0	27.1	54.1	87.1	121.8	170.1	216.6	243.6	270.7	284.2	284.2	284.2	284.2	284.2	284.2	277.3	270.7	270.7	270.7	209.7									
Monopterus 20	25.0	12.5	15.0			18.3	41.6	58.2	94.8	124.8	146.4	169.7	184.7	196.3	202.2	208.0	208.8	209.7													
LM 25 kW	25.0	15.0	na		7.5	16.6	28.1	41.2	55.6	71.6	90.7	112.1	132.9	143.8																	
Cauperduin	30.0	10.6	na		9.3	33.5	61.9	96.6	131.9	170.1	196.7	218.7	237.2	248.7	255.7	262.6	267.2	270.7													
12/30	30.0	12.0	14.0				36.1	63.2	99.3	126.4	162.5	198.6	234.7	266.3	252.8	248.2	243.7	234.7	230.2	225.7	225.7										
Aeroman 12.5/30	30.0	12.5	10.0			25.0	58.2	91.5	124.8	166.4	212.1	257.9	274.5	274.5	274.5	274.5	274.5	274.5	274.5	274.5	274.5	274.5	274.5	274.5	274.5	274.5	274.5	274.5	274.5	274.5
Darrieus DZ-12	30.0	12.0	18.0			19.6	67.7	108.3	135.4	198.6	243.7	270.8	288.9	291.1	293.4	288.9	279.8	275.3	274.4	273.5	272.6	271.7									
F.H.W. 30	30.0	12.5	15.0	14.1	25.0	41.6	63.2	91.5	127.3	168.1	221.3	252.9	252.9	252.9	252.9	252.9	252.9	252.9													
N 12/30	30.0	12.5	18.0			20.8	41.6	74.9	112.3	149.8	168.0	253.7	266.2	266.2	249.6	224.6	212.1	191.3	178.9	164.4	153.9	149.8	141.4	137.3	133.1	133.1	133.1	137.3	137.3	141.4	145.6
Kano-Rotor 30kW	30.0	12.9	15.5	8.4	18.0	28.7	47.8	71.7	110.0	145.8	179.3	210.4	236.8	251.1	261.8	265.4	263.0	255.9	253.5	254.7	264.3	277.4									
HSW-30	33.0	12.5	14.5		16.6	41.6	83.2	124.8	166.4	208.4	249.6	270.4	270.4	270.4	270.4	270.4	270.4	270.4	270.4												

Table 13: Specific power [W/m²] as a function of windspeed of small-scale WECs in the rating class 10.1 to 75 kW (contd.)
Data obtained from manufacturer's power curve

WEC-typname	rated power [kW]	rotor dian. [m]	tower height [m]	2.5	3.5	4.5	5.5	6.5	7.5	8.5	9.5	10.5	11.5	12.5	13.5	14.5	15.5	16.5	17.5	18.5	19.5	20.5	21.5	22.5	23.5	24.5	25.5	26.5	27.5	28.5	29.5
Aeroman14.8/33n	33.0	14.8	15.0	..	..	12.6	25.3	50.5	83.3	118.7	151.5	174.2	189.4	194.5	194.5	194.5	194.5	194.5	194.5	194.5	194.5	194.5	194.5	194.5	..	..	..	..	..	..	..
WPS11A40	40.0	11.5	17.0	..	..	16.0	33.2	52.9	81.1	118.9	160.2	201.5	241.8	285.0	329.3	361.2	393.2	..	..	..	..	..	..	..	..	..	..	..	..	..	..
N 1245	45.0	12.5	18.5	..	..	20.8	41.6	74.9	112.3	149.8	262.1	312.0	357.7	374.4	374.4	324.4	299.5	266.2	237.1	216.3	195.5	178.9	162.2	145.6	133.1	120.6	112.3	108.2	104.0	95.7	91.5
TW 45	45.0	12.5	na	..	..	4.2	20.8	49.9	83.2	123.1	166.4	224.6	274.5	324.4	366.1	407.6	420.1	432.6	428.4	424.3	428.4	435.1	..	..	..	..	..	..	..	..	..
BON 55/11	55.0	11.0	na	..	..	21.5	32.2	80.6	161.1	241.7	338.4	402.9	419.0	537.1	580.1	612.3	628.5	644.6	644.6	644.6	644.6	644.6	644.6	644.6	..	..	..	..	..	..	..
VEST 55/15.3	55.0	15.3	na	..	..	16.7	42.2	66.6	100.0	155.5	199.9	233.2	266.5	294.3	311.0	316.5	322.1	322.1	322.1	322.1	322.1	322.1	322.1	322.1	..	..	..	..	..	..	..
(55/16)	55.0	16.0	20.0	..	..	..	5.1	27.9	53.3	88.9	132.0	167.6	198.0	233.6	269.1	281.8	294.5	299.6	307.2	309.7	312.3	312.3	..	..	..	..	..	..	..	..	..
RET 55/15.6	55.0	15.6	na	..	..	10.7	26.7	53.4	80.1	117.5	186.9	240.4	277.8	293.8	309.8	301.8	293.8	283.8	267.1	251.1	245.7	235.0	..	..	..	..	..	..	..	..	..
WPS16A60	60.0	16.0	na	..	..	..	16.9	33.8	56.4	87.3	123.9	165.0	203.1	239.7	277.8	300.1	..	..	..	..	..	..	..	..	..	..	..	..	..	..	..
TW 60	60.0	16.9	23.5	..	..	9.1	34.1	58.3	91.0	118.3	147.9	182.1	207.1	227.6	245.8	259.4	264.0	254.9	259.4	266.2	266.2	..	..	..	..	..	..	..	..	..	..
M 66/13-U5	65.0	16.0	na	..	..	10.2	20.3	50.8	91.4	147.2	192.9	238.7	274.2	294.5	314.8	327.5	340.2	340.2	340.2	335.1	330.1	..	..	..	..	..	..	..	..	..	..
WR 65/17	65.0	17.0	na	..	..	4.5	9.0	15.7	24.7	38.2	54.0	72.0	101.2	134.9	179.9	224.9	258.6	292.4	292.4	292.4	292.4	..	..	..	..	..	..	..	..	..	..
15-75	75.0	15.2	22.5	..	..	16.9	56.3	101.3	163.2	213.8	264.4	315.1	365.7	410.7	450.1	472.6	495.1	512.0	523.2	534.5	540.1	542.9	545.7	548.6	551.4	..	..	..	..	..	..
VEST 75/15	75.0	15.0	na	..	..	28.9	57.8	104.0	144.4	196.4	236.9	288.9	323.5	369.7	404.4	427.5	450.6	468.0	479.5	479.5	479.5	479.5	479.5	479.5	479.5	..	..	..	..	..	..
NTK 75/16	75.0	16.0	na	..	..	15.2	40.6	67.0	107.7	151.3	208.2	253.9	288.4	309.2	330.1	343.2	356.4	365.1	373.2	377.3	381.8	384.6	385.9	..	..	..	..	..	..	..	..

na = data not available

For list of complete WEC-type names, see Appendix

Table 14: Specific power [W/m²] as a function of windspeed of WECs in the rating class 75.1 to 300 kW
Data obtained from manufacturer's power curve

WEC-typname	rated power [kW]	rotor diam. [m]	tower height [m]	2.5	3.5	4.5	5.5	6.5	7.5	8.5	9.5	10.5	11.5	12.5	13.5	14.5	15.5	16.5	17.5	18.5	19.5	20.5	21.5	22.5	23.5	24.5	25.5	26.5	27.5	28.5	29.5
EBERCON-17/80	80.0	17.2	28.0	3.9	13.3	25.4	43.0	74.2	113.3	164.0	214.8	275.4	320.3	345.7	351.5	351.5	351.5	351.5	351.5	351.5	351.5	351.5	351.5	351.5	351.5	351.5	:	:	:	:	:
Svedana 80/50Hz	80.0	17.0	24.0	13.5	31.5	45.0	63.0	76.5	99.0	134.9	179.9	247.4	278.9	296.9	310.4	328.3	350.8	346.3	350.8	346.3	337.3	328.3	323.9	319.4	319.4	328.3	328.3	337.3	341.8	346.3	:
UK T-1700	80.0	17.0	na	:	9.0	31.5	54.0	90.0	125.9	170.9	211.4	256.4	301.4	341.8	348.6	341.8	:	:	:	:	:	:	:	:	:	:	:	:	359.8	364.3	:
WINDANE 17/80	80.0	17.0	22.6	:	6.8	20.2	36.0	67.5	108.0	157.4	215.9	269.9	314.9	348.6	362.1	359.8	348.6	337.3	321.6	310.4	308.1	310.4	314.9	323.9	332.8	344.1	355.3	359.8	364.3	:
M080/18	80.0	18.0	na	2.0	:	6.0	12.0	22.1	44.1	63.4	96.3	112.3	144.4	196.6	278.8	299.9	321.0	321.0	:	:	:	:	:	:	:	:	:	:	:	:	:
15-90	90.0	15.6	22.5	:	6.9	13.3	26.7	48.1	98.8	141.6	186.9	248.4	296.4	333.8	363.2	383.2	403.3	416.6	422.0	427.3	427.3	422.0	416.6	:	:	:	:	:	:	:	:
NOAH90KW	90.0	17.0	na	:	2.2	11.2	27.0	45.0	66.6	90.0	121.4	148.4	175.4	197.9	220.4	227.1	233.9	231.6	229.4	231.6	235.2	:	:	:	:	:	:	:	:	:	:
AMP 90/18	90.0	18.5	22.5	:	:	12.7	29.1	60.0	90.9	125.5	174.6	225.4	272.7	305.7	330.9	345.5	361.8	363.6	360.0	354.6	341.8	327.3	327.3	330.9	:	:	:	:	:	:	:
V19-90	90.0	18.8	23.4	:	:	14.7	38.6	80.9	125.0	165.5	202.3	235.4	264.8	286.9	301.6	316.3	323.6	327.3	327.3	325.5	321.8	318.1	312.6	:	:	:	:	:	:	:	:
Svedana100/60Hz	95.0	17.0	24.0	4.5	18.0	36.0	58.5	85.5	116.9	170.9	242.9	283.4	332.8	364.3	404.8	416.1	427.3	422.8	413.8	409.3	404.8	395.8	395.8	395.8	409.3	413.8	418.3	422.8	:	:	:
WINDANE 19/95	95.0	19.0	22.6	:	:	7.2	32.4	64.8	104.4	158.4	208.8	257.5	295.3	324.1	338.5	342.1	334.9	327.7	316.9	306.1	298.9	300.7	302.5	309.9	300.7	315.1	322.3	331.3	338.5	:
MPS18A100	100.0	18.0	17.0	:	:	13.4	30.8	50.5	82.7	125.6	167.3	212.6	259.6	307.7	354.3	377.7	401.2	:	:	:	:	:	:	:	:	:	:	:	:	:	:
MPS18SW100	100.0	18.0	na	:	5.7	25.8	33.4	54.6	84.2	124.4	172.5	225.9	287.7	347.8	391.6	:	:	:	:	:	:	:	:	:	:	:	:	:	:	:	:
BON 100/19.4	100.0	19.4	23.5	:	:	13.8	31.1	56.1	94.3	150.9	200.3	254.7	300.5	329.8	353.7	371.1	388.6	386.8	378.9	380.6	388.6	388.6	388.6	388.6	:	:	:	:	:	:	:
BON 100/20	100.0	19.4	30.0	:	:	12.5	32.3	61.6	94.3	151.2	200.8	253.4	298.4	326.6	351.0	375.4	386.7	384.8	381.1	382.9	386.7	388.6	388.6	388.6	388.6	:	:	:	:	:	:
20-100	100.0	20.0	30.0	19.5	35.8	65.0	113.7	172.2	224.2	279.5	318.5	325.0	325.0	325.0	325.0	325.0	325.0	:	:	:	:	:	:	:	:	:	:	:	:	:	:
V 20-100kW	100.0	20.0	23.4	:	:	30.9	65.0	78.0	149.5	191.7	232.3	255.1	286.0	307.1	321.7	339.6	347.7	352.6	354.2	354.2	352.6	349.3	341.2	334.7	:	:	:	:	:	:	:
MPS 20	100.0	20.0	na	:	3.2	14.6	35.8	62.4	97.8	139.3	172.6	203.1	251.8	289.0	307.1	321.7	341.2	367.2	393.2	396.5	386.7	:	:	:	:	:	:	:	:	:	:
ADLER 25/100kW	100.0	25.0	22.0	:	8.3	18.7	35.4	62.4	97.8	139.3	172.6	197.6	208.0	208.0	208.0	208.0	208.0	:	:	:	:	:	:	:	:	:	:	:	:	:	:
M 100	108.0	19.3	na	:	:	7.0	17.4	41.9	83.8	139.6	202.4	258.2	303.6	335.0	362.9	375.1	387.4	390.9	390.9	390.9	390.9	:	:	:	:	:	:	:	:	:	:
M-19G0	110.0	19.6	23.0	:	:	3.7	17.3	51.1	96.4	149.9	209.1	273.4	321.4	354.6	379.3	393.4	407.4	411.5	415.2	416.2	416.2	416.2	416.2	:	:	:	:	:	:	:
NTK - 13MF	130.0	20.5	24.0	:	:	13.3	28.5	62.6	104.4	167.0	214.4	275.2	307.5	404.2	417.5	423.2	428.9	436.4	435.4	432.7	428.9	423.2	417.5	415.6	413.7	406.1	402.3	402.1	402.1	402.1
NTK 130/20.4	130.0	20.4	na	:	:	18.7	65.6	115.6	156.2	203.0	249.9	290.5	321.7	349.8	393.6	374.8	393.6	402.9	406.1	412.3	:	:	:	:	:	:	:	:	:	:	:
D 140	140.0	19.6	na	:	:	8.5	27.1	59.2	100.8	153.3	215.7	283.4	510.6	405.2	444.1	463.2	482.2	485.6	483.0	482.2	481.3	480.5	478.8	475.4	:	:	:	:	:	:

Table 14: Specific power [W/m²] as a function of windspeed of WECs in the rating class 75.1 to 300 kW (contd.)
Data obtained from manufacturer's power curve

specific power [W/m²] at windspeed [m/s]

WEC-typname	rated power [kW]	rotor diam. [m]	tower height [m]	2.5	3.5	4.5	5.5	6.5	7.5	8.5	9.5	10.5	11.5	12.5	13.5	14.5	15.5	16.5	17.5	18.5	19.5	20.5	21.5	22.5	23.5	24.5	25.5	26.5	27.5	28.5	29.5
DAMWIN 19/140	140.0	19.6	22.5			16.9	32.1	54.1	101.5	152.3	223.3	287.6	339.4	406.0	439.9	456.8	473.7	470.3	456.8	446.6	439.9	439.9	439.9	439.9	439.9	439.9					
20/150	150.0	20.0	30.0				29.2	52.0	87.7	133.2	204.7	266.5	338.0	409.5	477.7	500.4	494.0	487.5	484.2	484.2	481.0	481.0	481.0	481.0	481.0	481.0					
TW 150	150.0	20.5	na			6.2	40.2	80.4	126.8	185.6	269.1	352.6	414.5	470.1	482.5	476.3	471.7	467.1	482.5	488.7	491.8	491.8								
D 150	150.0	22.2	23.5			7.9	21.1	44.8	83.1	131.9	192.5	263.8	324.4	361.3	379.8	385.1	379.8	386.4	338.9	320.5	307.3	295.4	287.5	282.2	276.9	273.0					
V-S 150(V.2)	150.0	22.0	25.5		3.6	18.3	36.8	55.1	98.7	149.6	208.8	271.5	328.7	372.4	398.0	397.8	397.8	382.6	363.9	339.6	315.7	298.2	287.4	286.7	287.4					
Windmaster 150	150.0	21.8	21.8				32.8	49.2	71.1	93.0	123.1	153.2	191.5	232.5	320.0	391.1	410.3	410.3	410.3	410.3									
BONUS 150 kW	150.0	23.0	29.5		16.1	42.3	70.5	112.8	171.2	233.6	290.0	336.4	366.6	376.6	370.6	352.5	332.3	312.2	300.1	292.0	290.0	288.0	288.0	288.0	288.0	288.0	288.0	288.0	288.0	
V-S 150 (V.3)	150.0	23.0	30.5		3.3	16.7	33.7	50.4	90.3	136.9	191.1	248.4	300.8	340.7	364.2	364.0	363.8	350.0	333.0	310.7	288.9	272.9	262.9	262.3	262.9					
NTK 150 XLR	150.0	24.6	30.0			8.6	26.9	75.2	116.0	161.1	214.8	268.5	307.2	339.4	343.7	339.4	324.4	322.2	322.2	322.2	322.2	322.2	322.2	322.2	322.2					
Nordex 150 kW	150.0	27.0	30.0			12.5	33.0	69.5	122.1	183.7	230.9	258.6	275.5	285.3	288.0	288.0	284.4	279.1	274.6	271.9	276.4	284.4	295.1	311.1	331.7					
BOUMA 160/20	160.0	20.0	99.0				24.4	53.6	97.5	143.0	190.1	242.1	292.5	347.7	393.2	435.5	477.7	506.9	523.2	533.0	536.2								
LW 11	160.0	23.0	na			22.1	44.2	73.7	110.6	164.6	221.2	277.7	309.6	323.1	324.4	313.3	302.2	281.4	265.4	254.3	248.2	245.7	243.3	243.3	245.7					
Adler 25/165kW	165.0	25.0	22.0		8.3	18.7	35.4	62.4	97.8	139.3	201.7	226.7	274.5	314.1	339.0	341.1	343.2													
V-S 180	180.0	22.2	26.0				48.1	95.0	148.0	209.3	278.7	348.4	407.6	449.3	461.4	473.6	462.5	445.2	424.0	396.7	369.9	349.7	337.2	335.0					
Windmaster200kW	200.0	21.8	23.0				32.8	65.7	109.4	164.1	235.2	298.1	363.8	418.5	484.1	533.4	574.4	604.5	615.4	615.4	610.0	607.2	607.2						
Windmaster200kW	200.0	21.8	22.5				32.8	65.7	109.4	164.1	235.2	298.1	363.8	418.5	484.1	533.4	574.4	604.5	615.4	615.4	610.0	607.2	607.2						
200kW/d15m^2	200.0	23.0	23.6		7.4	19.7	36.9	61.4	98.3	140.1	189.2	262.9	314.5	407.9	486.5	489.0	491.4	491.4	491.4	491.4	491.4	491.4							
DAMWIN 24	200.0	24.0	29.0			12.2	36.8	73.1	119.4	173.3	232.4	294.3	351.8	399.9	437.4	448.9	431.5	402.8	345.2	352.1	341.4	337.2	344.6	357.7	369.9					
DAMWIN 24 (US)	200.0	24.0	23.0			12.2	36.8	73.1	119.4	173.3	232.4	294.3	351.8	399.9	437.4	452.9	448.9	431.5	402.8	345.2	352.1	341.4	337.2	344.6	357.7	369.9				
V25-200	200.0	25.0	28.7			15.2	36.4	67.6	108.2	151.8	199.7	251.7	292.2	355.6	400.4	416.0	416.0	416.0	416.0	416.0	416.0	416.0	416.0						
V-S 225	225.0	24.0	26.0				46.0	94.9	149.6	212.8	285.0	359.1	421.1	467.6	483.6	490.4	473.0	452.5	424.3	394.7	372.8	358.6	353.4						
Nordex 225/250	225.0	26.0	30.0			13.5	29.8	60.6	104.8	166.3	240.4	316.3	380.7	422.1	443.2	450.9	451.9	451.9	451.9	451.9	451.9	451.9	451.9	451.9					
AIT 02 - Medit	225.0	32.0	25.0				44.4	88.9	139.6	196.8	260.2	285.6	285.6	285.6	285.6	285.6	285.6	285.6	285.6								
TW 250	250.0	23.0	28.5		9.8	24.6	51.6	86.0	122.9	176.9	235.9	299.8	378.4	449.7	516.0	565.2	609.4	619.2	570.1										

Table 14: Specific power [W/m²] as a function of windspeed of WECs in the rating class 75.1 to 300 kW (contd.)
Data obtained from manufacturer's power curve

WEC-typname	rated power [kW]	rotor diam. [m]	tower height [m]	2.5	3.5	4.5	5.5	6.5	7.5	8.5	9.5	10.5	11.5	12.5	13.5	14.5	15.5	16.5	17.5	18.5	19.5	20.5	21.5	22.5	23.5	24.5	25.5	26.5	27.5	28.5	29.5
24.5m/250kW	250.0	24.5	25.0	..	..	..	28.1	71.5	119.1	173.2	231.7	298.9	357.3	413.6	463.4	508.9	539.2	550.0	554.4	554.4	554.4	554.4	554.4	554.4	554.4	554.4	554.4	554.4	554.4	554.4	..
M 450-250kW	250.0	24.0	28.7	..	..	11.3	31.6	63.2	119.6	180.5	255.0	320.5	385.9	453.6	519.0	564.2	577.7	575.5	561.9	541.6	512.3	485.2	..	..	..	..	..	..	..	..	..
HSW-250	250.0	25.0	27.3	..	58.2	104.0	135.2	197.6	270.4	364.0	426.4	468.0	503.3	536.6	551.1	613.5	584.4	594.8	603.1	605.2	607.3	605.2	..	..	..	..	..	..	..	..	..
NS-2	250.0	25.0	25.0	..	..	12.5	52.0	95.7	158.1	237.1	316.1	399.3	484.6	520.0	520.0	520.0	520.0	520.0	520.0	520.0	520.0	520.0	520.0	520.0	520.0	..	..	..	..	..
Windmaster 250	250.0	25.0	30.0	..	..	37.4	58.2	122.7	166.4	249.6	312.0	405.6	463.8	515.8	520.0	520.0	520.0	520.0	520.0	520.0	520.0	520.0	..	..	..	..	..	..	..	..
M 530-250kW	250.0	26.0	28.7	..	..	23.1	64.7	90.6	182.2	259.9	331.0	394.2	445.5	473.4	480.7	467.3	447.6	438.9	436.5	436.5	436.5	436.5	436.5	436.5	..	..	..	..	..	..
V-S 270	270.0	26.0	24.0	..	7.8	33.4	64.7	104.0	172.7	243.3	313.9	390.5	453.2	480.7	490.5	494.5	482.7	467.0	445.4	417.9	388.5	372.8	364.5	..	..	..	..	..	..	..
Windmaster 300	300.0	25.0	21.8	..	..	6.7	46.6	99.8	166.4	236.3	329.4	416.0	509.1	579.0	609.0	615.6	623.9	623.9	623.9	623.9	623.9	623.9	623.9	623.9	..	..	..	..	..	..
Nwmss-25	300.0	26.5	22.0	..	..	7.4	37.0	90.7	181.4	244.3	348.0	470.2	555.3	555.3	555.3	555.3	..	..	..	..	..	..	..	..	..	..	..	..	..	..
NTK 300	300.0	27.5	30.5	..	..	10.3	43.0	91.1	156.4	226.9	302.5	386.7	464.1	532.8	584.4	611.9	610.2	593.0	567.2	567.2	567.2	567.2	567.2	567.2	..	..	..	..	..	..
WPS30GSR300	300.0	30.0	17.0	..	10.1	31.8	56.3	85.2	124.2	174.8	238.3	317.8	398.6	436.2	..	..	..	..	..	..	..	..	..	..	..	..	..	..	..	..
WINDANE 31/300	300.0	30.8	30.8	..	13.7	41.1	75.4	123.3	178.1	253.5	335.7	397.4	411.1	411.1	411.1	411.1	411.1	411.1	411.1	411.1	411.1	..	..	..	..	..	..	..	..	..
Enercon 32/300	300.0	32.0	34.5	10.9	23.4	44.3	70.1	106.8	155.2	215.6	286.1	380.8	380.8	380.8	380.8	380.8	380.8	380.8	380.8	380.8	380.8	..	..	..	..	..	..	..	..	..
NS-3	300.0	33.0	25.0	..	..	14.3	41.8	77.6	125.3	185.0	268.6	358.1	358.1	358.1	358.1	358.1	358.1	358.1	358.1	358.1	358.1	..	..	..	..	..	..	..	..	..

na = data not available

For list of complete WEC-type names, see Appendix

Table 15: Specific power [W/m²] as a function of windspeed of WECs in the rating class 300.1 to 900 kW
Data obtained from manufacturer's power curve

WEC-typname	rated power [kW]	rotor diam. [m]	tower height [m]	4.5	5.5	6.5	7.5	8.5	9.5	10.5	11.5	12.5	13.5	14.5	15.5	16.5	17.5	18.5	19.5	20.5	21.5	22.5	23.5	24.5	25.5	26.5	27.5	28.5	29.5
WINDANE 34/400	400	34.8	30.0	10.7	42.9	96.6	139.5	201.8	268.3	332.7	397.1	429.3	429.3	429.3	429.3	429.3	429.3	429.3	429.3	429.3	429.3	429.3					
BONUS 450 kW	450	35.0	32.6	15.2	33.4	64.4	103.2	149.1	190.1	254.4	297.5	349.6	401.8	446.2	481.5	509.1	522.9	530.6	526.0	512.2	490.8	466.2	453.9	446.2	444.7	444.7	444.7		
FLODA 600	600	35.0	42.0	17.1	39.1	73.2	125.9	189.6	259.8	339.5	413.7	477.4	535.6	577.2	596.8	601.8	601.8	601.8	601.8	601.8	601.8	601.8	601.8	601.8	601.8	601.8	601.8	601.8	601.8
NIBE A	630	40.0	44.5	na	na	8.1	44.7	93.4	150.3	213.3	284.4	361.0	430.2	487.5	511.4	527.4	na	487.5										
NIBE B	630	40.0	44.5	na	12.2	50.8	93.4	150.3	211.2	284.4	365.6	446.8	483.4	487.5	487.5	487.5	487.5	487.5										
WINDANE 40/750	750	40.0	42.0	12.2	44.7	81.2	142.2	190.9	268.1	341.2	430.6	503.7	556.5	609.3	609.3	609.3	609.3	609.3	609.3	609.3	609.3	609.3					

Table 16: Specific power [W/m²] as a function of windspeed of WECs in the rating class > 900 kW
Data obtained from manufacturer's power curve

WEC-typname	rated power [kW]	rotor diam. [m]	tower height [m]	4.5	5.5	6.5	7.5	8.5	9.5	10.5	11.5	12.5	13.5	14.5	15.5	16.5	17.5	18.5	19.5	20.5	21.5	22.5	23.5	24.5	25.5	26.5	27.5	28.5	29.5
WKA 60	1200	60.0	50.0	na	21.7	54.2	93.9	140.8	213.0	288.9	375.5	433.3	433.3	433.3	433.3	433.3	433.3	433.3	433.3	433.3	433.3	433.3	433.3					
GAMMA 60	1500	60.0	61.6	na	43.3	72.2	90.3	135.4	180.5	270.8	361.1	451.4	541.6	541.6	541.6	541.6	541.6	541.6	541.6	541.6	541.6	541.6	541.6	541.6	541.6	541.6	541.6		
ESBJERG 1	2000	61.0	57.0	na	na	38.4	80.3	139.7	192.1	262.0	345.8	422.7	503.1	565.9	628.8	663.7	698.7	698.7	698.7	698.7	628.8	545.0	454.1	366.8					
LS-1	3000	60.0	45.0	na	na	68.6	151.7	241.9	321.4	422.5	523.6	642.7	758.3	888.2	1018.2	1083.2	1083.2	1083.2	1083.2	1083.2	1083.2	1083.2	1083.2	1083.2	1083.2	1083.2	1083.2		
GROWIAN	3000	100.4	100.0	na	9.7	48.4	96.7	145.1	203.1	275.6	348.2	386.9	386.9	386.9	386.9	386.9	386.9	386.9	386.9	386.9	386.9	386.9	386.9						

na = data not available

For list of complete WEC-type names, see Appendix

In Figures 16 - 19 the specific power curves of selected WECs in four rating ranges taken from table 12, 13, 14, 15 and 16 are plotted as a function of wind speed.

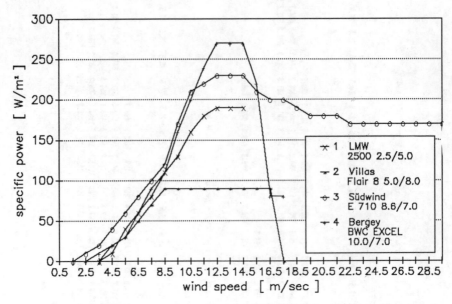

Fig. 16 :   Specific power of selected WECs with rating 0.1 - 10 kW as a function of wind speed

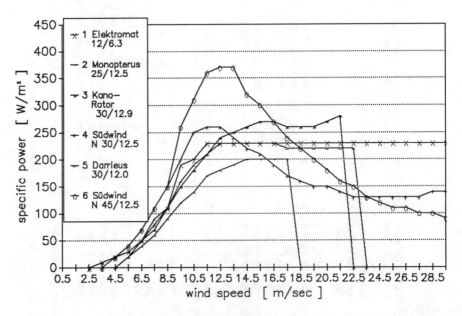

Fig. 17 :   Specific power of selected WECs with rating 10.1 - 50 kW as a function of wind speed

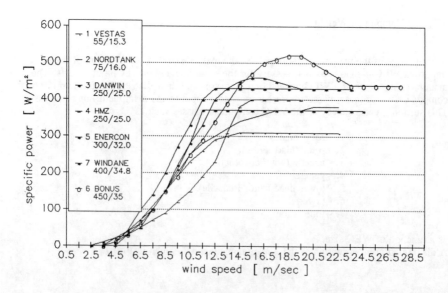

Fig. 18 : Specific power of selected WECs with rating 50.1 - 450 kW as a function
of wind speed

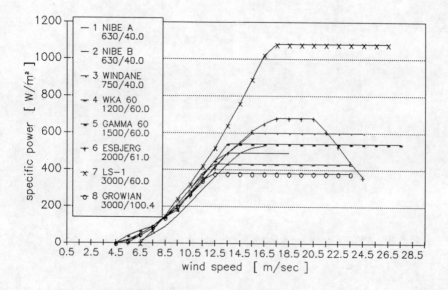

Fig. 19 : Specific power of selected WECs with rating > 450 kW as a function of
wind speed

5.4   Component Weights

A knowledge of the weights or more precise the masses and their distribution of the most important WEC components is necessary for the economic and energy input aspects. The manufacturing costs of a WEC depend on the masses and their distribution in the different components since these masses represenr an energy input to the manufacturinmg process. This section gives a survey of the WEC component masses such as

> **1.** - rotor blades
> **2.** - rotor (blades + hub)
> **3.** - nacelle (without rotor)
> **4.** - generator
> **5.** - tower head (rotor + nacelle)
> **6.** - tower
> **7.** - foundation

as a function of WEC size.
A diagramatic representation of these main components is given in Fig. 20.

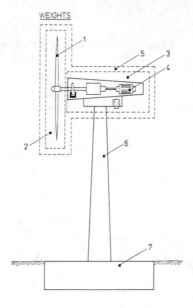

Fig. 20: WEC components selected for weight analysis

The following graphs show the statistical distribution of WEC components as a function of WEC size in two different ways. For each component the first figure inidicates the absolute weight in [kg] or [tons] and is often plotted on a logarithmic scale. The second figure shows the specific weight of a particular component, expressed as its weight divided by the rotor swept area or rated power in [kg/m²] or [kg/kW].

First, Figure 21 shows the weight of rotor blades, according to blade material, as a function of the blade length.

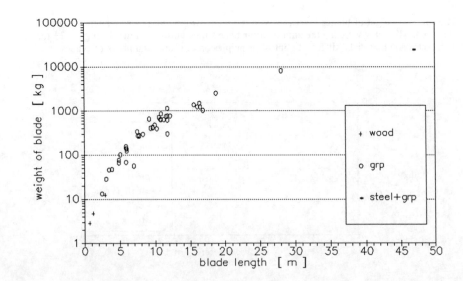

Fig. 21 : Blade weight for different blade materials as a function of blade length

Based on the diagram above, the specific weight distribution of rotor blades can be indicated as in Fig. 22.

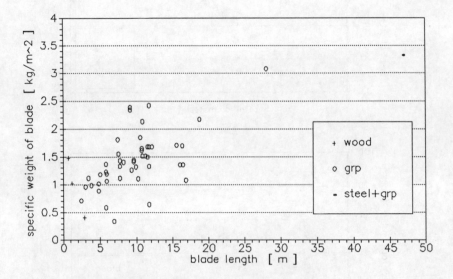

Fig. 22:   Specific weight of rotor blades for different blade materials as a function of blade length

The specific weight of blades increases with the blade length but a clear tendency is not observed. The rotor weight - the sum of rotor blades plus hub - is shown in Figure 23 for different blade materials, different control principles and different numbers of blades.

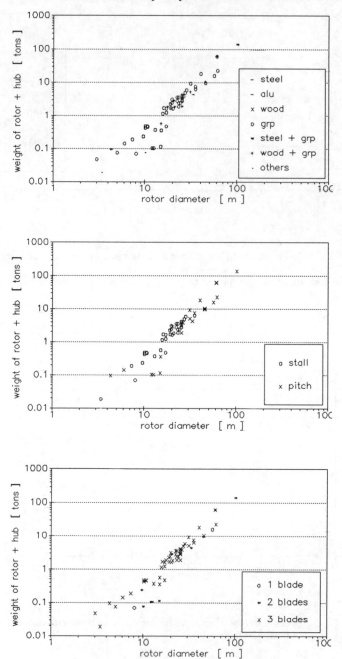

Fig. 23 :   Rotor weight according to blade material, control principle, and number of blades as a function of rotor diameter

The diagram in Fig. 24 shows the corresponding specific weight of the rotor, also according to blade material, control principle and number of blades.

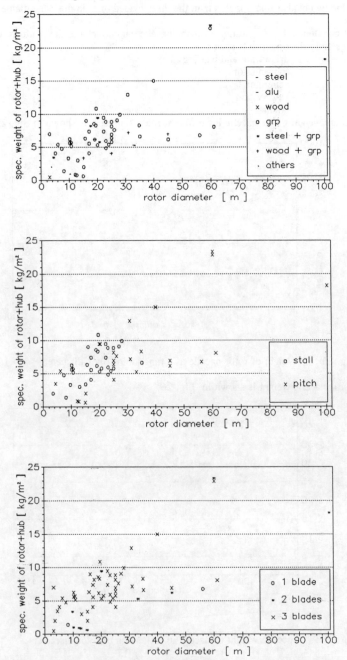

Fig. 24 :   Specific weight of the rotor including hub according to blade material, control principle, and number of blades as a function of rotor diameter

The specific weight increases with rotor diameter and, as expected, the majority are three bladed rotors consisting of glasfibre reinforced plastic rotor blades. The specific weight of most of the two bladed rotors is less than the three bladed ones in the same diameter category.

The next component to be investigated is the nacelle which includes the platform with all mountings and sub-components such as

- drive train and bearings
- transmission
- generator
- yaw system
- bonnet

The absolute weight of the nacelle for different rotor diameters is shown in Fig. 25.

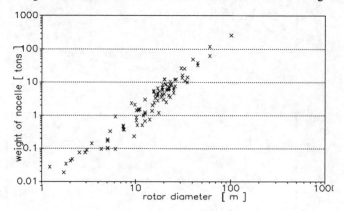

Fig. 25 :   Weight of rotor nacelle as a function of rotor diameter

The specific nacelle weight is shown in Fig. 26.

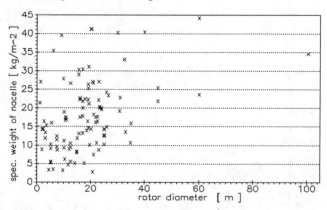

Fig. 26 :   Specific weight of nacelle as a function of rotor diameter

A factor 10 is observed in the spread of the specific nacelle weight, which means that the design of the nacelle is not primarily determined by aerodynamic forces and weight loads. The generator is not a specific wind energy component. Its weight is plotted as a function of rated power in Figure 27.

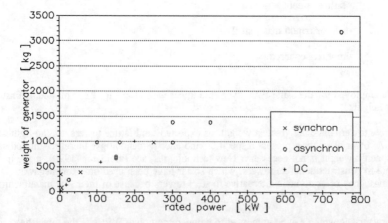

Fig. 27 :   Weight of generator according to generator type as a function of its rated power

There is a nearly linear relation between generator weight and rated power, and there is no significant difference between synchronuos, asynchronous or dc generators.

The last single component of this series is the tower, whose weight distributions are shown in Fig. 28.

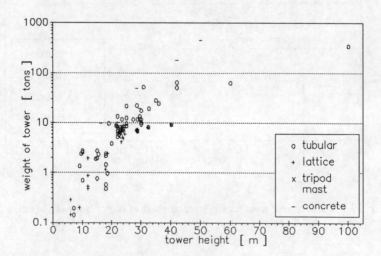

Fig. 28 : Tower weight as a function of tower height

Because the different types of towers, as well as various heights and construction materials such as

- tubular, steel

- lattice, steel

- trio or tripod mast, steel

- tubular, concrete

can be employed for the same WEC type, the tower weight in Fig. 28 is plotted against the tower height.

Concrete towers have the greatest weight, as expected, and lattice towers are the lightest ones. A tower does not represent a specific wind energy technology, therefore the indication of a specific weight is not necessary. New technologies, see Pfleiderer (1990), make it feasible to construct concrete towers, which so far have been used only for megawatt machines; they offer a viable alternative to steel towers because of their low manufacturing costs.

A characteristic which can be employed to compare complete WECs is the tower head mass. This mass as a function of rotor diameter is shown in Figure 29, followed by the specific tower head mass in Fig. 30.

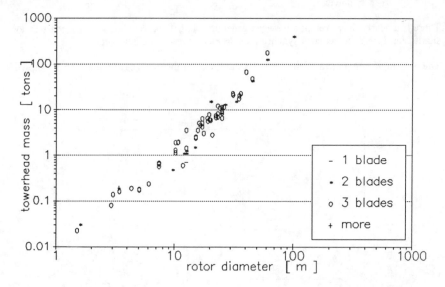

Fig. 29 :   Tower head mass for different numbers of rotor blades as a function of rotor diameter

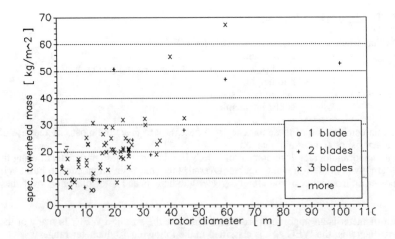

Fig. 30 :   Specific tower head mass for different numbers of blades as a function of
             rotor diameter

Although there is a spread with a factor 5, the specific tower head mass increases with rotor
diameter. The majority of machines are three -bladed and as a general rule the two-bladed
machines have a lower specific tower head mass than the three-bladed ones.
WECs with more than three blades appear only in the range less than 5 m rotor diameter.
To complete this series of diagrams, the total weight of WECs including tower head mass
plus tower mass is shown in Fig. 31 on a logarithmic scale on the vertical axis.

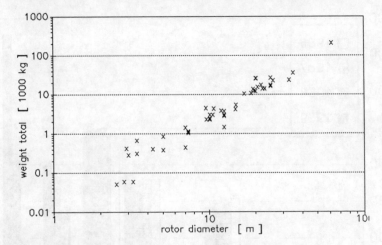

Fig. 31 :   Total weights of WECs including tower head and tower mass as a function
             of rotor diameter .

5.5   Trends in WEC Characteristics

The most important design quantities of a modern WEC are
- rotor diameter,
- rated power, specific rated power and power curve
- hub height,
- design wind speeds
- control principle

An optimum combination of these parameters is necessary to make the most efficient use of the wind characteristics at a certain location.

Trends in the technological development of WECs in relation to these parameters during the period 1986-89 are given in the following figures. Overall, there has been a tendency to build WECs with larger power ratings and correspondingly greater hub heights and rotor diameters over these four years.

The rotor diameter is an important parameter to indicate the size of a WEC. In many of the following diagrams, the WEC size is classified in the following 10 diameter ranges :

1:	0 - <6.3m	6:	14 m - <16 m
2:	6.3m - <8 m	7:	16 m - <18 m
3:	8 m - <10 m	8:	18 m - <20 m
4:	10 m - <12 m	9:	20 m - <31 m
5:	12 m - <14 m	10:	>31 m

It is interesting to note the different rates of increase for different rotor sizes, shown in Figure 32 for 10 different rotor diameter classes over 4 years. The figure also shows the trend toward the installation of larger, more powerful WECs in these years.

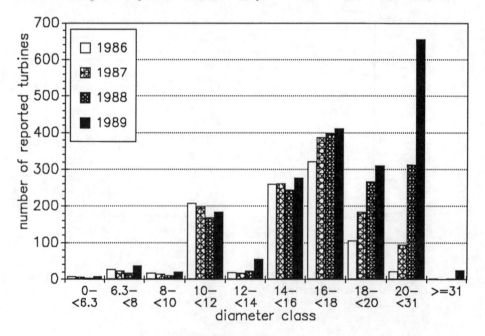

Fig. 32 : Distribution of reported WECS for different rotor diameter classes and years

Reported WECs with rotor diameters less than 10 m are relatively rare. In the rotor diameter classes of less than 10 m, the WECs consist mostly of self-constructed turbines which often work in the stand-alone mode (ie. not connected to a central grid). As a policy for the EUROWIN database, self-constructed WECs with a rated power of less than 10kW are excluded. This may explain the relatively low reporting quote for WECs with rotor diameters less than 10 m.

It should be noted that while the reporting quotes for WECs with rotor diameters between 10 and 16 m remained relatively constant or even decreased slightly, the reporting quotes for WECs in the medium-high rotor diameter classification (between 16 and 30 m), increased at a high rate. WECs with rotor diameters greater than 16 metres have been the most popular for the years between 1986 and 1989. The reporting rates for these classes increased from year to year and the highest rate of increase was observed in the class 20 - <31 metres, i.e. from 20 in 1986 to 650 in 1989.

The rotor diameter is closely related to the installed capacity of a wind turbine. The reported installed capacity of all WECs is defined as the sum of the rated power of all WECs reported. In Fig. 33 it is plotted against the rotor diameter class.

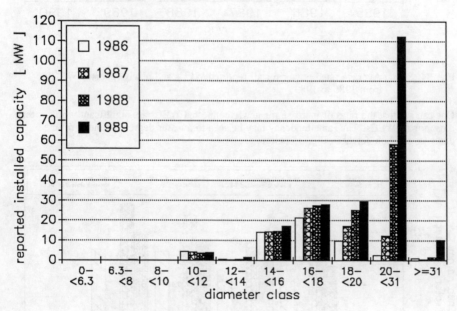

Fig. 33 :   Installed capacity of reported WECs for 10 rotor diameter classes between
               1986 and 1989

This graph also shows that most of the installed wind power capacity is from WECs with rotor diameters between 16 and 30 m. In general, while the reported installed capacity for WECs with a rotor diameter less than 10 m has decreased between the years 1986 and 1989, the installed capacity for WECs with rotor diameters in the medium-high classification (16-30 m) has increased in that period.

The average rated power for all WECs and the newly connected WECs for each year are plotted in Figure 34 for the years 1986 - 1989. The average rated power has increased for newly reported WEC's each year since 1986. This again indicates the continual trend towards increased operation of larger WECs.

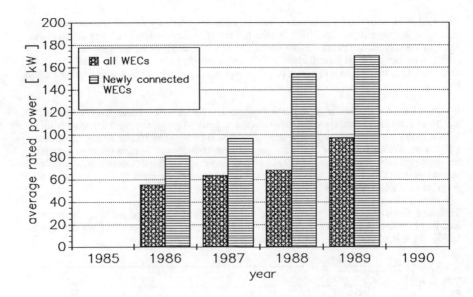

Fig. 34 :   Average rated power for all reported WECs and newly reported WECs
            from 1986 to 1989

One can see that the rates of increase for newly connected WECs are significantly higher than
those for all WECs. The same tendency may be seen to a lesser degree for mean rotor
diameter and hub height in Fig. 35 below.

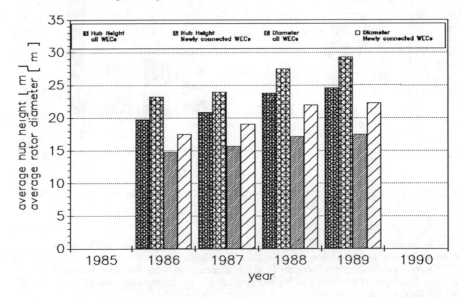

Fig. 35 :   Average rotor diameter and hub height for all reported WECs
                                   between 1986 and 1989

Figure 36, the last in this series, shows the development of the average specific rated power.

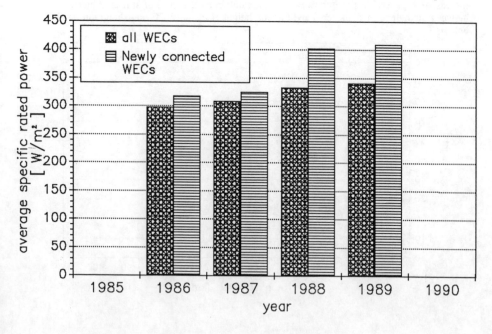

Fig. 36 : Mean specific rated power for 1986 - 1989

Wind speed increases as a function of the height above ground. Therefore, by increasing the hub height of the WEC, advantage can be taken of higher wind speeds. Consequently more power can be extracted from the wind, and the specific rated power of the WEC can be expected to increase. This increase in the average rated power, the hub height, and the rotor diameter can be seen by comparing Figures 34, 35 and 36.

The four Figures which follow are statistics relating to the design of individual WEC types since 1986. Some of them are no longer being produced by the manufacturers or have already been dismantled (GROWIAN). However, for statistical comparison of their technical data they still apppear in the diagrams below.

In the database WECs are distinguished by their control principle, such as
- stall   full blade stall
- wing tips full blade stall, moving wing tips
- flaps   full blade stall, flaps
- pitch   full blade pitch
                   - electric/hydraulic
                   - mechanical
- flex   flexible blade
- yawing
- variable geometry

The diagrams below merely distinguish between stall and pitch control. In brief, for the pitch control strategy, the power of the WEC is regulated by changing the angle of attack of the rotor blades. In stall controlled WECs, the pitch angle of the rotor blades remains fixed and the power of the rotor is controlled aerodynamically by the stall effect.

Obviously as the hub height of the WEC is increased, the diameter of the rotor can also be increased. A general practice, however, is to maintain a difference of at least 2 meters between the half rotor diameter d and the hub height H of the WEC, see the graph in Fig. 37 below.

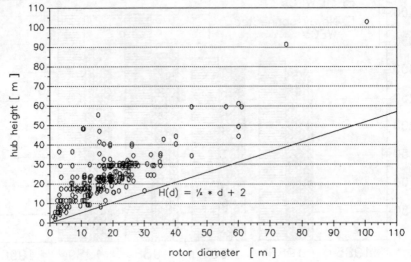

Fig. 37 : Hub height as a function of rotor diameter for individual WEC types

Figure 37 illustrates the linearity between increasing hub height and increasing rotor diameter. The choice of tower height is an optimization between increasing energy gain and manufacturing costs with tower height.

How far the increase in energy gain with height above ground really leads to higher annual energy generation will be treated in detail in section 6.1.6.

Fig. 38 graphically displays the relationship between the rated power and the rotor diameter of various registered WECs. The graph is plotted on a double logarithmic scale.

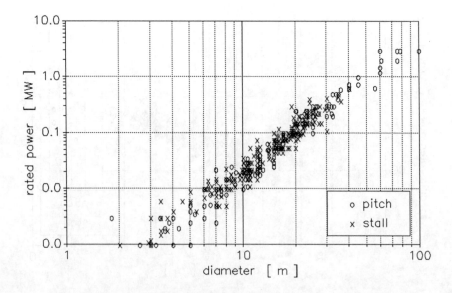

Fig. 38 :   Installed rated power vs. rotor diameter for selected types of WECs
                registered in EUROWIN

The specific power curves for various WECs have already been discussed in section 5.3. The
specific power of a WEC is only one point in this characteristic and is obtained from the
rated power at rated wind speed divided by the rotor swept area.
The performance of WECs is strongly dependent on the variability of the wind, and
deviations from the optimal wind conditions lead to losses in the efficiencies of WECs.
The driving design of WECs is characterized by quantities such as
                - specific rated power
                - design wind speeds

Designs are optimized for certain wind site characteristics to receive a optimum energy
generation.
These design conditions are different for stand-alone and grid-connected WECs. In general,
stand-alone wind turbines are designed to supply a high load demand, while grid-connected
WECs are designed and optimized for high annual energy production.

The specific rated power of several existing WECs is examined in detail in the next three
figures. In general, a low potential wind energy is coupled with a low specific power rating
whereas a high mean annual wind speed may lead to WECs with higher specific power
ratings. Fig. 39 shows the specific rated power for various WECs as a function of the WEC
size, characterized by rotor diameter and rated power.

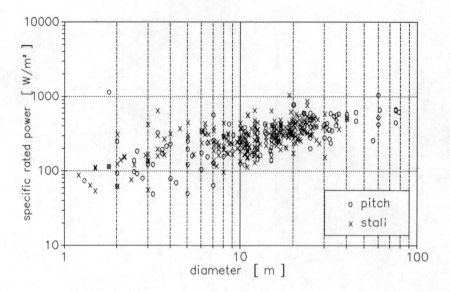

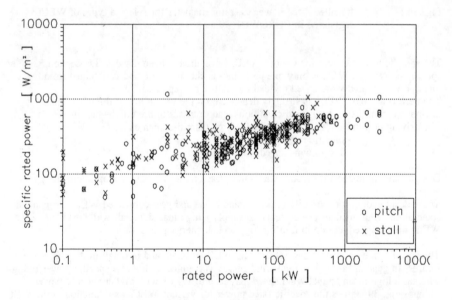

Fig. 39: Specific rated power as a a function of WEC size for selected types of WECs

No difference is evident for the two different control strategies of stall and pitch control but the slope flattens at higher rated powers.

Figure 40 below is a graph of the specific rated power of the different types of WECs plotted against their rated windspeed. The upper curve is the power density of the wind as a function of the wind speed of the free flow v_0 and the lower curve indicates the corresponding Betz limit.

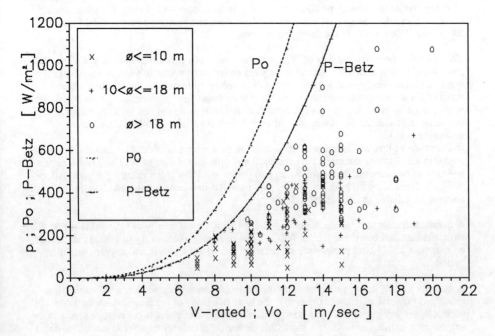

Fig. 40 : Specific rated power vs rated wind speed of selected WECs

In general, the WECs with smaller rotor diameters do not follow the Betz limit power curve as closely as the WECs in the larger diameter classes. The whole of the specific power characteristic of a WEC (see sect. 5.3) which includes the point of specific rated power in Fig. 40, must be located beneath the p-Betz curve. Otherwise it would exceed the physical limit set by the Betz maximum.

6	Evaluation of Operational Data 1986-1989

The analysis in this section is compiled from the operational data in the EUROWIN database. Monthly and annual energy production, capacity factors, operational time, failure analysis, economics and the energy payback time will be treated and presented in graphical form for the period 1986 - 1989 in the following sections.

The analysis of operational data of WECs can be seen from two different points of view. The first consideration is that the natural wind yield changes from year to year. A knowledge of this fluctuating energy over long time periods is necessary to answer the question : How much of the conventional generated electricity from fossil-fired and nuclear power plants can be displaced by wind energy use ? Secondly, a statistical evaluation of WEC operational data can show the technical developments and improvements which took place during the evaluation period.
In the frame of this work, few wind measurement data were available. The data obtained and stored in the database comprise only technical and operational data of WECs. Therefore one question to be examined in the following paragraphs is : How is the energy generation of WECs influenced by high or low energy gain potential in a certain year, and how it is affected by reliable and advanced technology ?

A question discussed since the beginning of commercial wind energy use is : What is the most efficient and economic size of WECs ? The following sections try to provide an answer by means of a statistical comparison of differently sized WECs over a four year evaluation period.

A quantity usually employed to quantify the quality of energy conversion within a physical process is the total efficiency. Generally, the total efficiency of machines which convert chemical, mechanical or electrical energy increases with size because the relative losses of the sub-components decrease. Over a short time period the efficiency of a WEC can be indicated by the power coefficient of the rotor, see equ. (3). For long-term comparisons the physical efficiency is more difficult to determine because a simultaneous measure of wind speed and power output is not available for commercially operating WECs.
To allow a comparison of different-sized WECs, the specific energy, which is the absolute energy generated in a certain time period divided by the rotor swept area in [kWh/m² time], is a central quantity in WEC evaluation. The energy generated and the capacity factor of various individual WECs throughout Europe are good measures of comparison between wind energy and other methods of energy generation. They are treated in detail in sections 6.1 and 6.2 under various criteria.

6.1   Energy Generation

6.1.1   Energy Generation 1986-1989

Figure 41 shows the absolute monthly energy production in [GWh] for all the WECs reported each month over four years. As shown by the dotted line in this graph, the number of turbines steadily increased between the years 1986 and 1989, (the gap in September 1989 means that there was an unusually low reporting quote). Correspondingly, the average energy generated by the WECs has increased each year.

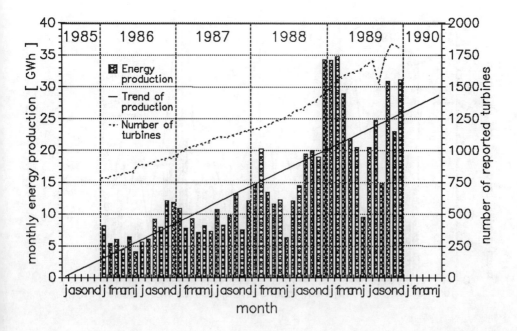

Fig. 41 :   Absolute monthly energy production and number of reported turbines
between January 1986 and December 1989

The linear graph in this figure indicates the trend function, which is based on the
assumptions, that the energy generation is proportional to the number of WECs and that the
increase of reported WECs is nearly linear.

The wind speeds vary both in a short-term pattern (seconds) and in a long-term pattern
(months). During the winter months the wind speeds are highest. This is the time when the
most energy can be generated from the wind. During the summer months (eg. June, July) the
wind speed falls. This is the time when the least power is available from the wind. This
yearly trend is noticeable in the graph of Figure 41. The peaks in the GWh per month appear
in the months of December and January and in June and July the least energy is produced.

In the Figure 42, the mean monthly specific energy for all reported WECs and the
corresponding capacity factor, which is defined and examined in detail in section 6.2, from
all reported WECs are shown between January 1986 and December 1989.

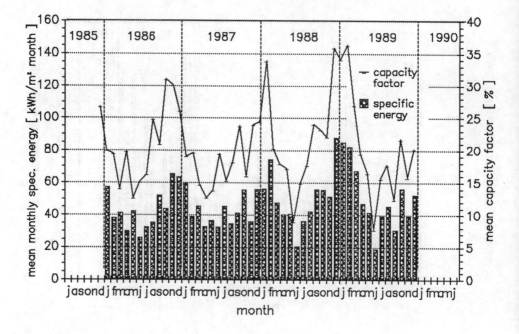

Fig. 42: Mean monthly specific energy and capacity factor between January 1986 and December 1989

A look at the course of the capacity factor over four years show, that the peaks and minima tended to become more extreme from year to year. On the one hand, more summer lulls with a low wind yield, on the other hand stronger gales in the winter months. The summer lulls, at any given year in June, and the winter maxima, always located between November and January, reached the following values of mean capacity factor :

mean capacity factor :

year		summer lull	winter peak
1986	:	13 [%]	27 [%]
1987	:	13 [%]	31 [%]
1988	:	9 [%]	34 [%]
1989	:	8 [%]	36 [%]

The maximum 1989/1990 is not included in this statistic but everyone remembers the four gales which raged over Europe within six weeks in January to February 1990, a singular event in this century.

The last figure in this series, Fig. 43, shows the mean monthly specific energy for all reported WECs in each year since 1986.

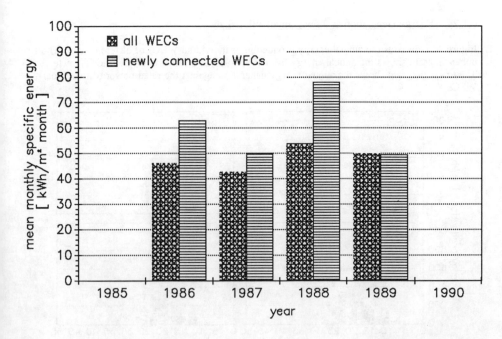

Fig. 43: Mean monthly specific energy for the total and newly connected
         WECs for 1986 - 1989

Because newly connected WECs usually have less than 12 reported months of data during the
first year of operation, the mean monthly instead of annual specific energy is a better measure
for comparison. New machines have significantly higher rates of specific energy than the
averages over all WECs.

Note that the mean monthly specific energy decreased in 1987. It is highly unlikely that all
WECs would decline in performance at the same time unless there were some external
causes. The decrease in specific energy for the year 1987 indicates that this was a bad wind
year. In 1989 most of the newly connected WECs began operation in the second half of the
year. This period was also a relatively bad wind period, see Fig. 40, which explains the
relatively low values for 1989.

6.1.2   Energy Generation as a Function of WEC Size

Figure 44 compares the annual energy production of individual WECs of different sizes. The upper diagram shows the annual energy from 1986-1989 as a function of the rotor diameter. The lower diagram shows the annual energy generation against the rated power of individual WECs.

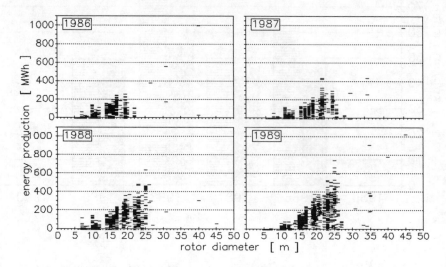

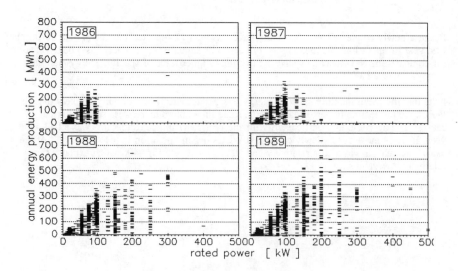

Fig. 44 :   Annual energy produced by individual WECs as a function of WEC size from 1986 - 1989

As regards energy generation, the upper diagram shows that the best machines follow a quadratic function with increasing diameter, while in the lower diagram, a linear relation between energy generation and rated power is observed. This is because the energy generation varies directly with the power, see equ.(9), whereas the power varies with the square of the rotor diameter.

The number WECs reported to the EUROWIN database has increased each year, especially for WECs with rotor diameters between 18 and 25 m. The WECs with rotor diameters of 40 m are the two Danish wind generators NIBE A and NIBE B (NIBE B appears only in the data for 1986).

In Figure 45 the specific annual energy for individual WECs is shown for the years 1986 to 1989 as a function of rotor diameter and rated power. The specific annual energy is defined as the annual energy divided by the rotor swept area or rated power for each WEC. The upper diagram shows the specific energy per rotor swept area in [kWh/m² a] whereas the lower diagram the annual energy generation of each WEC is divided by its rated power which leads to [kWh/kW a].
Both diagrams are distinguished to stall- and pitch-controlled machines.
The upper diagram shows an increase of specific energy until rotor diameters of 25 m. In the diagram below the dependency of specific energy from the rated power [kWh/kW a] is not as obvious as from the rotor swept area in the diagram above [kWh/m² a].

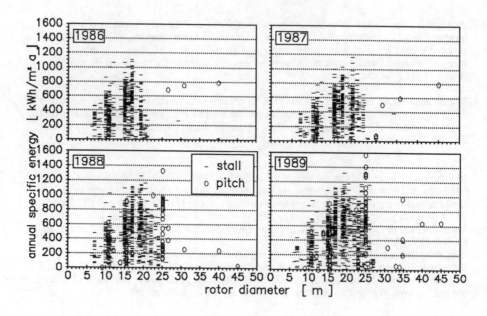

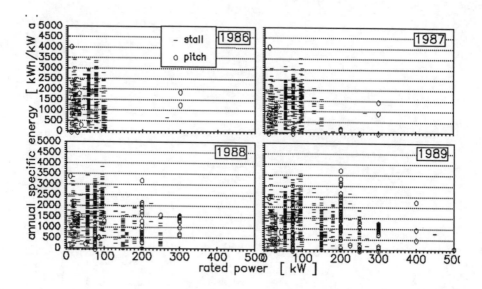

Fig. 45 : Annual specific energy generation as a function of WEC size 1986 - 1989, distinguished for stall and pitch controlled machines

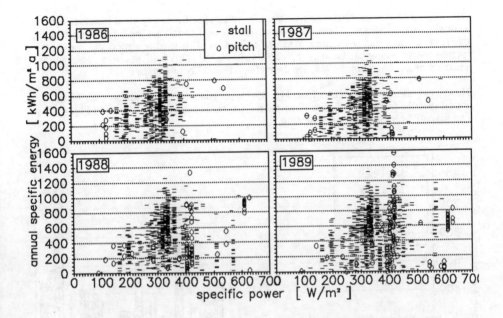

Fig. 46 : Annual specific energy generation as a function of specific rated power
from 1986 to 1989, distinguished for stall and pitch controlled machines

Fig. 46 shows that the majority of the newer machines appear with higher specific ratings
from year to year. There is a trend towards higher specific energy with increasing installed
specific power. In Fig. 35 it was shown that the hub height also increased continuously and
therefore higher installed ratings are possible because of higher wind speeds. Since 1989 the
majority of the best machines are represented by pitch controlled machines with a specific
rating approaching 400 W/m².

From Figure 45 it is interesting to note that the annual specific energy yield increases with increasing rotor diameter. Above 20 m diameter it appears to fall but the few machines in this size range are new and not mature. When an efficient WEC is used at a good site, the specific energy output can be more than 1000 [kWh/m2 a]. This level has been exceeded for a number of machines since 1988, see the table of TOP TEN in section 6.1.4.

The data of all individual WECs from Fig. 45 can be summarized to express their mean annual specific energy in ten rotor diameter classes, see section 5.5, as shown in Fig. 47. The years indicate the year of evaluation.

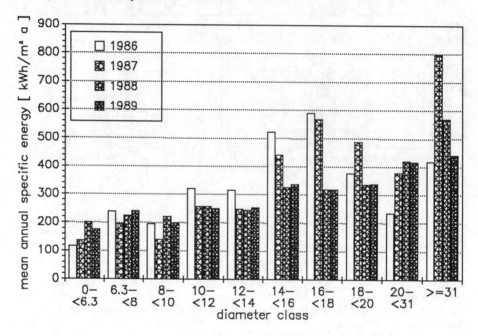

Fig. 47 :   Mean annual specific energy for ten diameter classes
             evaluated between 1986 and 1989

Since 1986 bigger WECs have been built, thus contributing to an increased yearly energy output from 1986 to 1989. A continuous increase of specific energy is observed with increasing rotor diameter classes.

This figure should be observed in conjunction with Figure 32 which plots the number of different WECs as a function of their rotor diameter class for the years 1986 - 1989. Each bar in the histogram of Fig. 47 must be 'weighted' with the corresponding number of evaluated WECs presented in Fig. 32. In the diameter class > =31m the bars in Fig. 47 do not represent a real statistic because only a few WECs are presented in this class.

6.1.3   Bin-Frequency of Specific Energy

The data of Fig. 45 can also be used to investigate the frequency of certain bins of specific annual energy. The specific energy is classified in bins of 100 [kWh/m² a] in Fig. 48.

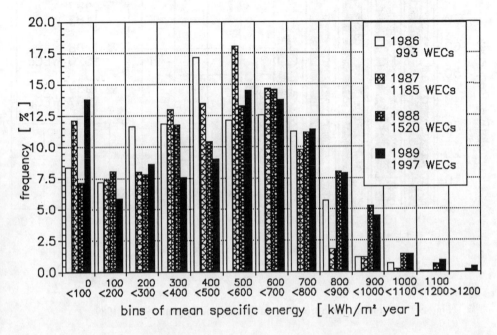

Fig. 48 :   Bin-frequency of specific energy from 1986 to 1989

The single bars show the frequency of the average specific energy in each bin, obtained from the data in Fig. 45.

The position of the maxima in eachyear can be located

in 1986 at 400-500 (avg.450) [kWh/m² a],
in 1987 at 500-600 (avg.546) [kWh/m² a],
in 1988 at 600-700 (avg.653) [kWh/m² a],
in 1989 at 500-600 (avg.530) [kWh/m² a]

In the higher bins, which represent the better machines, an increase of frequency is observed from year to year with the exception of 1987. This reflects the improvements in wind energy technology.

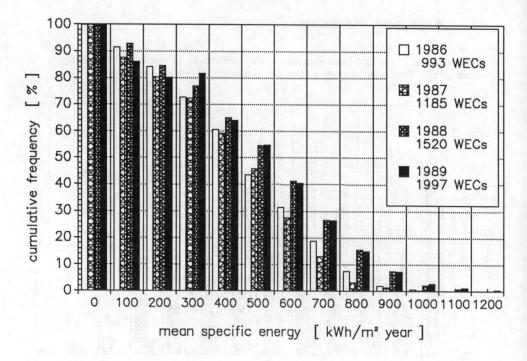

Fig. 49 :   Cumulative bin-frequency of specific energy from 1986 to 1989

The cumulative frequency distribution corresponding to Fig. 48 is shown in Fig. 49. Since 1988 a significant increase in frequency is observed for all bins above 400 kWh/m².

The best machines can be found in the upper bins above 1000 kWh/m² year, represented by approximately 2% of all machines. A detailed analysis of the TOP-TEN is the theme of the following section.

6.1.4   Top Ten Ranking WECs for Energy Production 1986-1989

The specific energy is a good indication of operating efficiency among WECs with different power ratings and rotor diameters. Table 17 below is lists the WECs with the highest annual specific energy output for the years 1986 to 1989. Note that although the larger WECs generally have the highest specific energy ratings, it is not necessarily true that the biggest WECs are always the best. Excellent results in energy production are achieved by a combination of proven technology with best wind conditions.

In Table 17 the trend towards constructing increasingly larger WECs is obvious. In 1986, the top ten WECs listed have a rated power between 45 and 75 kW, and in 1987 the rated power is between 55 and 95 kW. In 1988 the best machines have a rated power between 75 and 200 kW and in 1989 between 95 and 200 kW.

Similarly, each year the top ten WECs produce increasingly higher specific energies.

Table 17 : Top ten rated WECs with highest specific energies per year for 1986 - 1989

Ranking order of WEC	1986 [kWh/m² a] [kW]/[m]	1987 [kWh/m² a] [kW]/[m]	1988 [kWh/m² a] [kW]/[m]	1989 [kWh/m²] [kW]/[m]
1	1120 75/17	1177 95/19.4	1343 200/25	1579 200/25
2	1096 75/17	1099 95/19.5	1271 95/19.4	1418 200/25
3	1083 55/15	1041 75/17	1193 150/23	1334 95/19.6
4	1081 45/12	1020 75/15	1171 130/20.5	1329 150/23
5	1036 75/17	995 95/19.4	1163 150/23.5	1319 200/25
6	1026 75/16	953 99/20.5	1149 95/19.4	1308 150/23
7	1021 75/17	947 55/15	1137 95/19.5	1305 200/25
8	1017 55/11	940 95/19.4	1127 80/17	1285 200/25
9	993 75/17	939 80/17	1114 75/17	1282 130/19.4
10	964 55/16	933 75/17	1106 95/19.5	1190 130/20.5

To clarify the presentation of Table 17 : for each WEC the first line gives the specific annual energy, and the second line the rated power and rotor diameter.

Example : WEC 1, 1986 : 1120 [kWh/m² a]; 75 kW; 17m

6.1.5   Country Comparison and Monthly Energy Index

The yearly profile of the mean specific monthly energy is shown in Figure 50 for three countries : Denmark, The Netherlands and Germany in the year 1989. Yearly specific energy profile is similar for each country. In the summer months of May and June, the specific monthly energy output is at its minimum. In the winter months of January and February, the maximum mean monthly specific energy is produced.

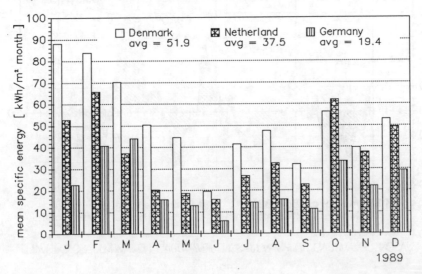

Fig. 50:   Monthly mean specific energy for Denmark, The Netherlands and Germany for the year 1989

The trend of Fig. 50 is similar to the other years as shown in Fig. 42. For the period 1986-1989 (see Table 18) the mean specific monthly energy is best for Denmark at 48.5 [kWh/m² month] followed by The Netherlands and then Germany with an for year average of 24.5 [kWh/m² month]. This is a significantly lower value. In part, this may be due to the fact that Denmark and The Netherlands have had more WECs in operation for a longer time than Germany, where the experiences with commercial wind energy use were just beginning. This table should be viewed in conjunction with Table 1, which gives the number of reported WECs for each country.
The overall averages, which are weighted with the number of turbines for each plant, are indicated in Table 18 below:

Table 18 :   Overall averages of specific energy for Denmark, The Netherlands, Germany, and Europe

	Denmark [kWh/m²month]	Netherland [kWh/m²month]	Germany [kWh/m²month]	Europe [kWh/m²month]
1986	45.3	31.5	xxx	44.5
1987	42.6	24.4	xxx	41.8
1988	51.5	50.3	32.5	50.9
1989	51.9	37.5	19.4	50.1
avg.	48.5	38.5	24.5	47.6

The averages over four years from January 1986 to December 1989 in the bottom row in Table 16 may be taken as a monthly energy index which can be calculated for Europe (all WECs reported in Europe) and for each country (all WECS reported in one country). Based on this monthly energy index, the monthly deviations of specific energy from January 1986 to December 1989 can be indicated, see Fig. 51-53. These figures also give graphs of the number of WECs with reported energy production in the same period.

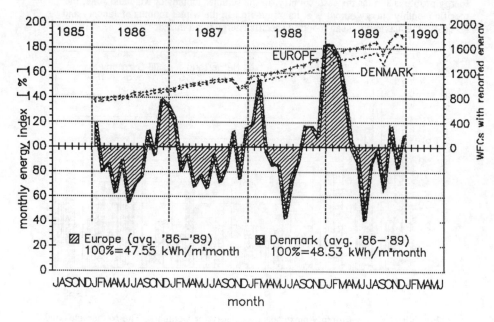

Fig. 51 : Monthly energy index for Europe and Denmark from 1986 to 1989

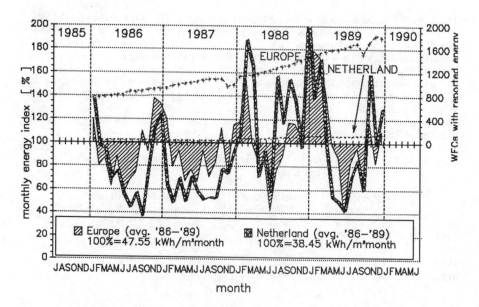

Fig. 52 : Monthly energy index for Europe and the Netherlands from 1986 to 1989

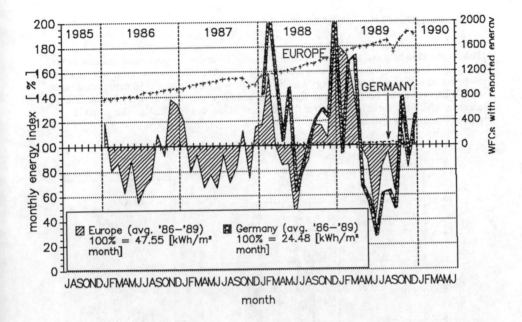

Fig. 53 : Monthly energy index for Europe and Germany from 1986 to 1989

Fig. 51 illustrates the dominant influence of Denmark within the evaluated period. There is almost no deviation between the European and Danish profile, also the two curves which indicate the number of monthly reported WECs are very close together.

Although in The Netherlands the number of reported WECs is much smaller and the overall average is lower, the qualitative trend is similar to the European index.

The same remark applies to Germany, where the number of monthly reported WECs never exceeded 60 but the qualitative trend with its characteristic peaks and troughs is similar.

Using Figures 51 to 53 it is possible for each operator of a WEC to compare the monthly results of his windturbine with those from Europe or another country. It should be borne in mind that one has to consider the overall average indicated by 100% in the legend of each figure.

6.1.6 Relationship of Hub Height to Specific Energy

The relationship between the hub height and the specific energy is interesting because as the hub height increases, the rotor blades are exposed to higher wind speeds. However, in Figure 54 below, the annual specific energy for WECs with hub heights above 30 metres decreases. Few WECs have hub heights greater than 35 m, and these machines have annual specific energies which are significantly lower than the best of the smaller WECs. The graph is parametrized to the 3 diameter classes.

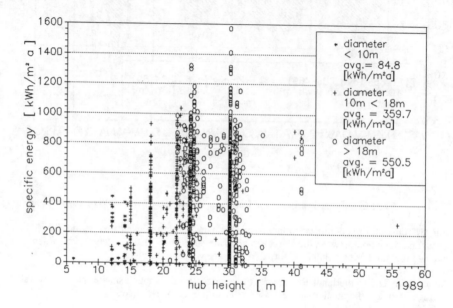

Fig. 54:   Specific annual energy vs hub height for various rotor diameters classes in the year 1989

In Fig. 55 the annual specific energy is plotted as a function of the relatio from hub height to rotor diameter H/d.

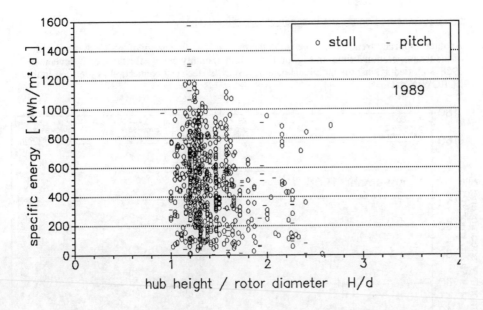

Fig. 55 :   Specific annual energy vs relatio hub height to rotor diameter
               in the year 1989

While in Fig. 54 the specific energy increased with the absolute hub height, in Fig. 55 no
clear tendency is observed for the relatio H/d.

6.1.7   Specific Energy as a Function of Operational Time

Another important trend to observe is the increase in specific energy of WECs as a function
of the total operational time. In Fig. 56 the specific monthly energy for the month november
1988 is plotted against the individual date of a WEC's electrical connection as reported in the
EUROWIN database.

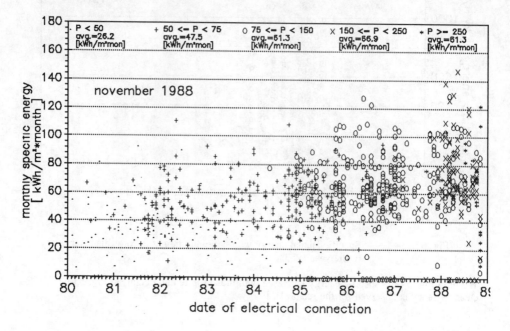

Fig. 56:   Monthly specific energy in November 1988 vs the individual date of a
            WEC's electrical connection

The reported WECs are divided into five different classes of rated power which received the
following average values in specific energy :

1.   0.1   -   < 50 kW avg.: 26.2 [kWh/m² a]
2.   50   -   < 75 kW   " : 47.5   "
3.   75   -   < 150 kW   " : 61.3   "
4.   150   -   < 250 kW   " : 66.9   "
5.   > 250 kW   " : 61.3   "

The month of November 1988 has been chosen as a representative example because over four
years the monthly energy index in November has been close to 100%, see Fig. 51. Similar
trends can be found in other months or for the cumulative mean annual specific energy, see
Fig. 56.

Figure 56 shows that until 1985, most of the WECs installed had a power rating of less than
75 kW. Since the beginning of 1986, most WECs had ratings higher than 75 kW and since
the beginning of 1988, the majority of newly connected turbines had ratings higher than 150
kW.

From this figure, it is seen that the newly connected WECs statistically have greater monthly specific energies than older WECs. This statement is verified by Figure 57.

Figure 57 shows the mean annual specific energy of all reported WECs as afunction of their year of electrical connection.

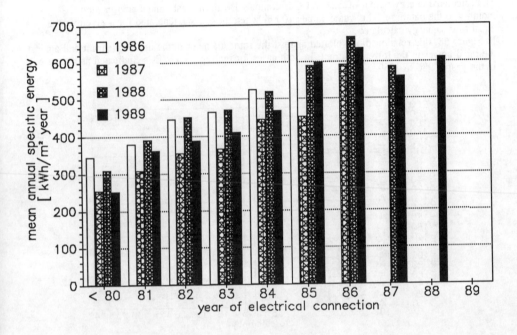

Fig. 57 : Mean annual specific energy as a function of the year of electrical connection

Each single bar in Fig. 57 represents one year of evaluation. For example the white bars, always located at the left, represent 1986 as the year of evaluation. Likewise the darkest bars, always located at the right, represent 1989 as the year of evaluation.

The newer machines, connected at the end of the 80's nearly reach almost twice the mean specific energy of WECs connected at the beginning of the 80's. Together with Fig. 55 this proves that while the WEC size increased continuously in the recent years an improvement in WEC technology also took place, see also Schmid; Klein (1990).

In the year of electrical connection which represents the beginning of operation, WECs usually have less than 12 reported months. This leads to a low cumulative annual energy and for that reason an indication about this quantity has been omitted from Fig. 57.

6.1.8   Reported and Calculated Energy Output

The annual energy generation of a WEC placed on a certain site, can be calculated by superposition of its power curve and the wind speed profile specific to the site location, see Fig. 6. For many WECs registered in the EUROWIN database, this expected annual energy production has been made available by the manufacturer or by a national wind energy institution such as Riso/Denmark, ECN-Petten/Netherlands or German Lloyd/Germany.

The calculated energy output of each WEC is compared with the real annual energy reported monthly to the database. This value equals 100 % when the energy generated for a given year is equal to the energy calculated.

In Figure 58, this relationship is plotted against the rotor diameter in the upper quadrant and against the rated power in the lower quadrant, in both cases for the reported WECs (indicated as measured) in the years 1986 to 1989.

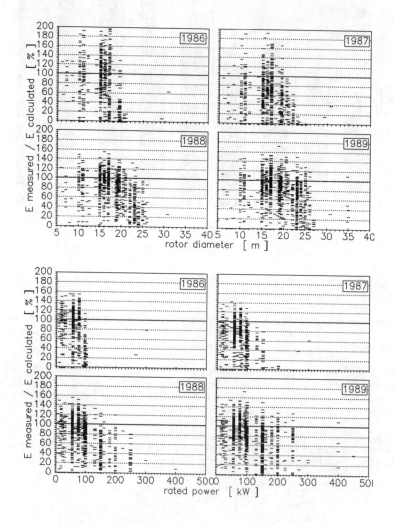

Fig. 58 :   Ratio of reported to calculated annual energy production vs WEC size
             from 1986 to 1989

A conclusion over four years and ten rotor diameter classes is shown in Fig. 59.

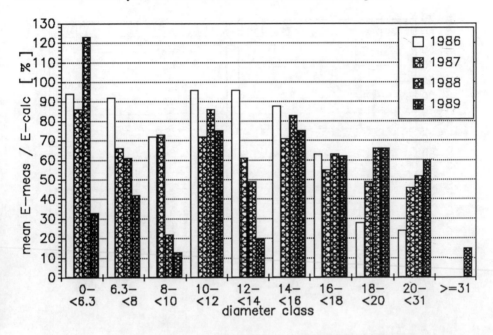

Fig. 59 :   Comparison between measured and calculated annual energy generation as a
function of ten rotor diameter classes

The individual WECs in Fig. 58 show large deviations from the calculated annual energy. The
overall average for all evaluated machines from Fig. 59, which is important for evaluating the
contribution of wind energy to the total electric energy supply, is given in Table 19.

Table 19 :   Average conformity of all reported WECs between real and calculated annual
energy generation

year	$[E_{meas.}/E_{calc.}]_{mean}$
1986 :	86.3 [%]
1987 :	71.8 [%]
1988 :	79.1 [%]
1989 :	77.7 [%]

In future, when more wind measurement data become available, it will be feasible to establish a
European wind index.

6.1.9 Energy Generation per Tower Head Mass

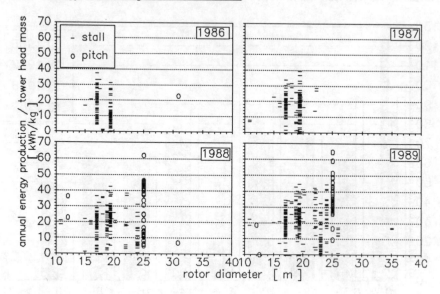

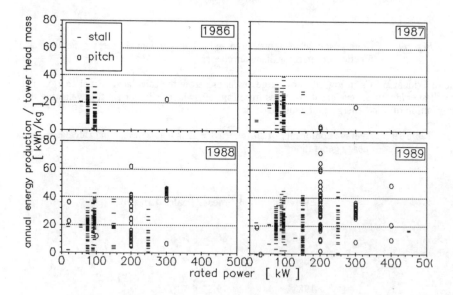

Fig. 60 : Energy generation per tower head mass as a function of WEC size between 1986 and 1989

A clear trend in the relation of annual energy to tower head mass cannot be observed. Since 1989 number of evaluable machines increased and the pitch controlled machines above 25 m diameter or 200 kW rated power reached higher values than the stall controlled machines.

6.2   Capacity Factor

The capacity factor of a WEC is an indication of the technical quality and reliability of a WEC as well as an indication of the adaptability of the WEC to the local wind speed conditions on site. This factor is the percentage of rated power received in a certain time period and is defined as the actual amount of energy produced in a given time period (E_T) divided by the rated power (P_{rating}) integrated over that same time period (T), see figure 61 below.

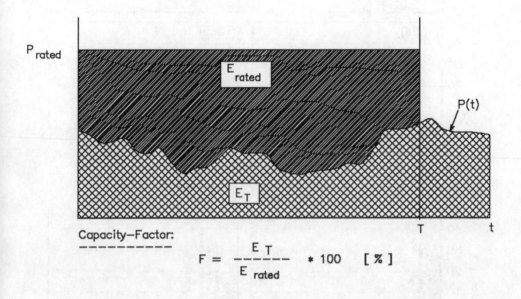

fig. 61 : Definition of capacity (or load) factor and mean power

In figure 42, section 6.1.1, the mean monthly capacity factor and specific energy were already graphed between 1986 - 1989 as a function in time history. This graph illustrates the yearly pattern followed by both specific energy and capacity factor. The profile exhibits a a peak output in the winter months and a minimum output in the summer months each year. It shows the close correspondence between wind speed variations over the year and the specific energy output and capacity factor.

In most cases, the capacity factor is defined over a period of one month or one year (8760 hours.). Figure 62 shows the annual capacity factor plotted against the WEC-size.

6.2.1   Capacity Factor as a Function of WEC Size

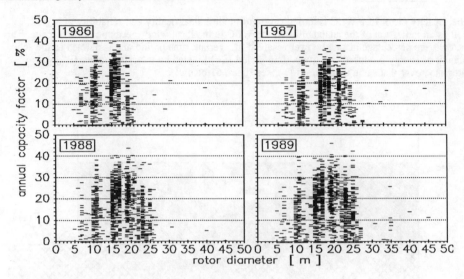

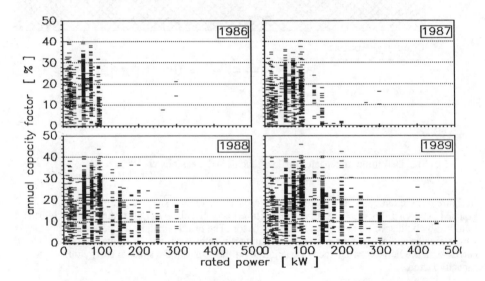

fig. 62 :   Annual capacity factor of reported WECs as a function of WEC-size for
the years 1986-1989

Realistic annual capacity factors of good WECs are usually between 25 - 30 %, but at times the
capacity factor of the best WECs as regards technology and siting has reached more than 40%. It is
worth noting that some manufacturers tend to rate machines in a rather arbitrary manner. Often
national subsidy arrangements cause machines to be given high ratings which leads to lower
capacity factors. For capacity credits, public utilities often focus on the capacity factors; by
decreasing the rated power in relation to the site characteristiscs, capacity factorsare increased.

The last figure in this chapter represents the average capacity factors of 10 diameter classes for all reported WECs between 1986 and 1989.

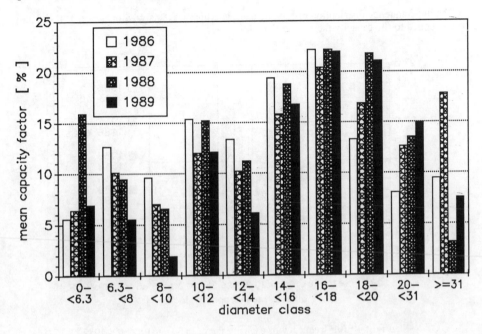

fig. 63 :   Mean capacity factor of all reported WECs as function of rotor diameter class for the years 1986 - 1989

One can observe the yearly increase of average capacity factors with a maximum in the range 16-20m between 1986 and 1989.

6.2.2   Bin-Frequency of Capacity Factor

The bin-frequency distribution of specific energy, which was analysed in sect. 6.1.3, can also be applied to the capacity factor, see Fig.64.

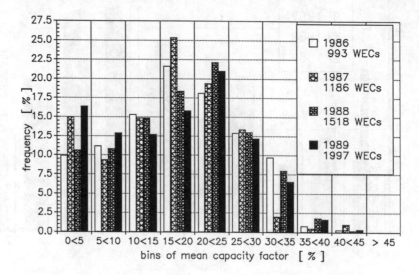

Fig. 64 : Bin-frequency of capacity factor between 1986 and 1989

The maximum frequency, which represents the majority of reported WECs, was within the 15-20 [%] bin in 1986/1987 and shifted towards 20-25 % in 1988 and 1989.

6.2.3 Capacity Factor as a Function of Operational Time

To allow for the fact, that newly connected WECs usually have less than 12 reported months in the year of their electrical connection, Fig. 65 shows the mean monthly capacity factor as a function of the year of electrical connection.

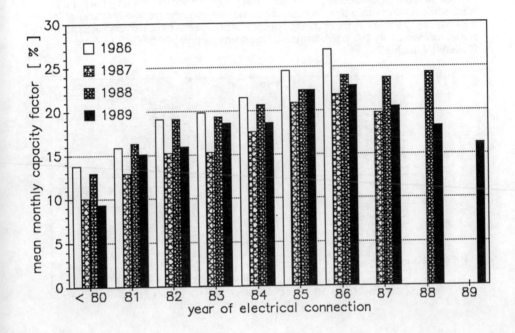

Fig. 65 :   Mean monthly capacity factor as afunction of the year of electrical connection

The years in Fig. 65 indicate the year of evaluation. As in Fig. 57, the dark bars at the right represent 1989 as the year of evaluation of the capacity factor. For example, in 1989 all WECs connected in 1981 had an average capacity factor of 15%. In the same year those WECs with 1986 as year of electrical connection reached about 23% of the mean capacity factor.

Because the capacity factor is a quantity derived from energy generation and rated power, this figure leads to the same conclusion as Fig. 45 regarding the specific energy as a function of operational time. For all four curves, a continuous increase or improvement from year to year is observed for the mean monthly capacity factor.

6.3 Operational Time

6.3.1 Operational Time as a Function of WEC Size

The operational time of a WEC is defined as the running time when a WEC is supplying energy to the load or grid. The availability additionally includes those periods when a WEC is not in operation because of low wind speeds. The operational time can be quantified by different methods. Some operators may individually approximate the operating time of the WEC while at other sites a timer is used. In general, the operators are mainly interested in the kWh output and not in the operating time of a WEC. Due to these inconsistencies, this section can serve only as a rough analysis of the operating time of the WECs reported to the EUROWIN database.

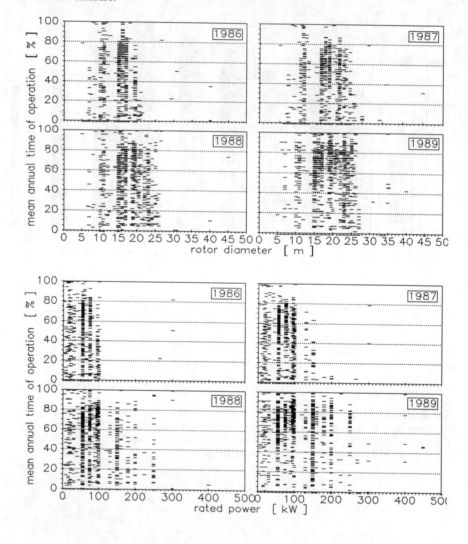

Fig. 66 :   Mean annual operational time of all reported WECs as a function of WEC size for the years 1986 - 1989

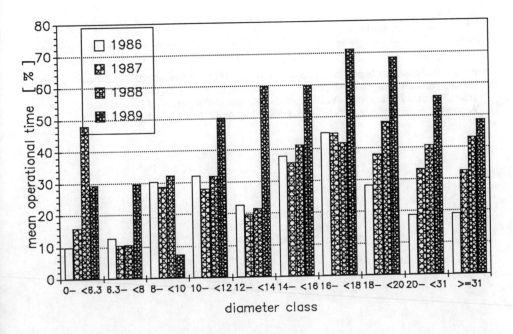

Fig. 67 :   Mean operational time of all reported WECs as a function of the
rotor diameter class for the years 1986 - 1989

Figure 66 shows the annual operational time of individual WECs as a function of WEC size.
The high operational time of over 80 % may be explained by the fact that many operators
used an individual approximation and not a timer to determine this parameter.

In Figure 67 the mean operational time for WECs in different diameter classes is plotted for
the years 1986-1989.

7   Failure Analysis

Failure analysis is an important and complex area in WEC evaluation. A WEC contains many
moving parts and thus the problems such as noise, large vibrations, and mechanical failures
arise. The monthly report to the EUROWIN database includes a list of failures, see
Appendix. Each operator reports failures according to the following categories:

1. **Control System**
2. **Yaw System**
3. **Rotor Blades**
4. **Generator**
5. **Grid**
6. **Mechanical Brakes**
7. **Gearbox**
8. **unspecified Failures**

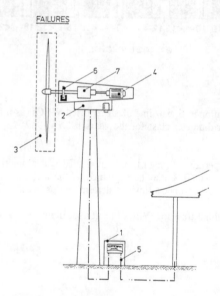

Fig. 68 : Evaluated faulty WEC components

7.1   Percentage and Relative Failures

Reported failures are analysed in two ways, namely : percentage failures and relative failures. A monthly percentage failure is defined as the sum of all observed failures per month divided by the number of reported WECs each month, see Fig. 69. A relative failure is defined as ratio of all observed failures within a ceratin period, see Fig. 69, or for a certain component, see Fig. 71, to the total number of failures observed during the month or year.

Figure 69 shows the percentage and relative failures between 1986 and 1989.

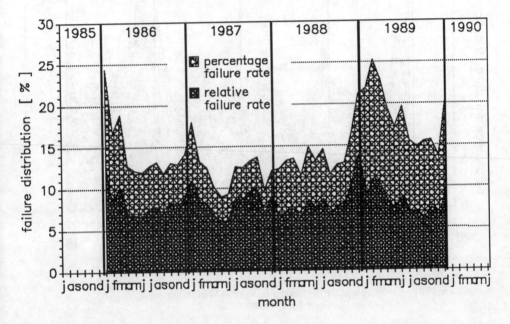

Fig. 69 :   Monthly percentage and relative failure distribution
for the years 1986 - 1989

One can observe a similar pattern for both failure distributions. In the winter months there is a higher failure quote than in the summer months. This may be due to the higher wind speeds and the adverse environmental conditions such as snow, ice, and rain , which are present in the winter.

7.2   Component Failure Analysis

Figure 70 compares the percentage failures of different components from 1986 to 1989. The component categories 1-8 are as shown in Fig. 67.

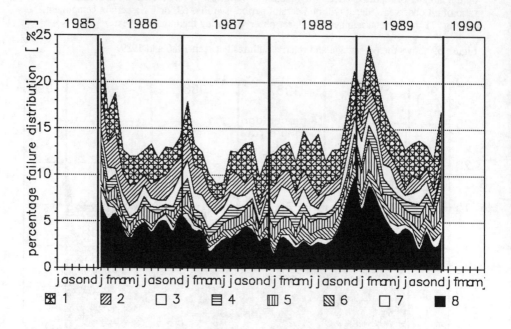

Fig. 70 :   Monthly percentage failures of WEC components
                   for the years 1986 - 1989

Of the recorded failures, most occurred in the control system and yaw system. 3 - 8 % of the reported WECs had failures which were in the 'unspecified' category. This indicates either that the cause of failure was unknown or that the failure occurred in a component other than those specified. The reasons for failures in this category should be investigated in more detail in future.

Figure 71 gives the relative distribution of the failed components for 1986 to 1988.

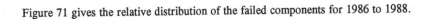

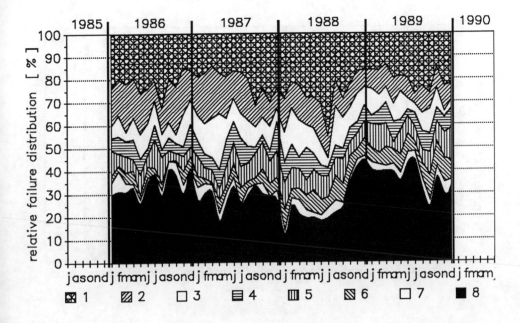

Fig. 71 :   Relative failure distribution of different failured components for the
                years 1986 - 1989

7.3   Classification of Failures

All failures that have been reported to the EUROWIN database are classified into three
groups as followis :

1.   Small Failures
          Failures which do not entail more than 24 hours repair timr

2.   Big Failures
          Failures which need between 24 and 72 hours repair time

3.   Total Breakdown
          Failures which lead to the total destruction of the WEC

Figure 72 indicates the total number of WEC failures reported as a function of time, and Fig. 73 shows the same percentage distribution of the three failure groups.

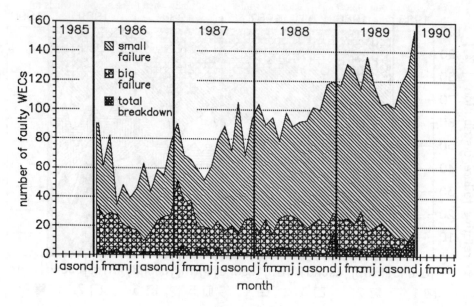

Fig. 72 :   Number of faulty WECs reported as a function of time

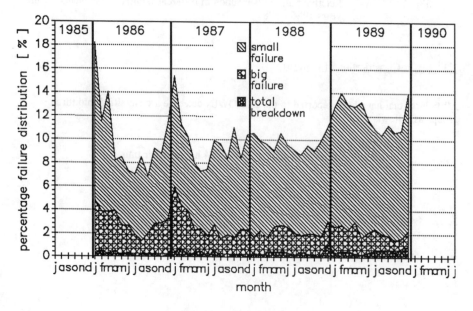

Fig. 73 :   Percentage failure distribution as a function of time

Each year has its own pattern of failures. It may be noted, however, that the course of the small failure distribution is similar to that of figure 42, which shows the yearly profile of the specific energy and the capacity factor. In winter and in times when there are high winds or harsh weather, the probability of failures will of course increase.

7.4   Operational Time and Failure Rate

The average component failures per WEC are recorded below in Fig.74 as a function of the year of electrical connection. The indicated years represent the year of evaluation. Note that each WEC can have several failures per year, so that the average failure per unit can be greater than 1 per year.

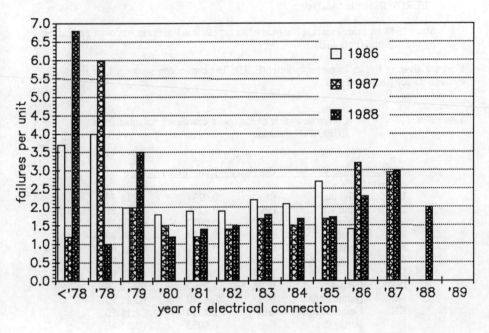

Fig. 74 :   Average failure rate per WEC as a function of the year of electrical connection   (1977-1988)

For each year evaluated, it can be seen that the failure quote is relatively high in the two years after the start of operation. After experience has been gained with the newly installed WEC, the failure rate decreases in the following years and then remains nearly constant. After seven years operation, the failure rate again increases. This pattern of failures is typical of many industrial products.

8	Economics of Wind Energy

8.1   Absolute and Specific Costs

There has been much discussion of the question of the most economical size of a WEC. The investigations and results presented here are a statistical answer which shows the development of the specific costs and generation costs. The indications of costs originated from the following sources :

- Institut for Vedvarende Energi
- Folkecenter
- Landwirtschaftskammer Schleswig Holstein
- Interressenverband Windkraft Binnenland
- EUROWIN market analysis   (*)

(*)   Since 1988 an annual market analysis of European wind turbines has been carried out by the institute (FhG-ISE)

Cost are given in European currency units (ECU). The conversion rates used for the evaluations are given in Table 20.

Table 20 :   Conversion table of European currencies obtained from the EUROWIN database

Abbr.	Country	Currency	Rate (ECU)	Date
community :				
B	Belgium	BFr	0.0231	08.11.89
DK	Denmark	DKr	0.1246	08.11.89
E	Spain	Pta	0.0077	08.11.89
F	France	FFr	0.1427	08.11.89
FRG	Germany	DM	0.4831	08.11.89
GR	Greece	GDr	0.5760	30.11.89
I	Italy	Lire	0.0007	08.11.89
IRL	Ireland	Ir.P	1.2590	08.11.89
NL	Netherlands	HFl	0.4270	08.11.89
UK	United Kingdom	GBP	1.4830	08.11.89
P	Portugal	Esc	0.0053	08.11.89
LUX	Luxembourg	LFr	0.0231	08.11.89
external :				
AU	Austria	ÖS	0.0697	14.03.90
CH	Switzerland	SFr	0.5560	08.11.89
J	Japan	Y		
S	Sweden	SKr	0.1425	08.11.89
USA	US of America	US $	0.9420	08.11.89

Figure 74 shows the total costs in [ECU] as a function of WEC size, parametrized to the year of census.

These costs are defined as the total costs for the erection of a WEC by side of its manufacturer, such as
- costs ex works
  + foundation
  + transport
  + erection

But for an operator who wishes to install a WEC there are additional costs which differ greatly in individual cases, such as :
- electric grid connection, depending from local infrastructure and generator size
- consultancy,
- land costs

These additional costs do **not** appear in Fig. 75.

The prices in Fig. 75 are **not** corrected for inflation and therefore indicate the real prices in the year of census. The inflation rates differ from country to country and could not be made available to the data base. Further investigation will be necessary to find out the development of costs as a function of time.
The total costs follow a course which is nearly quadratic to the rotor diameter and is linear to the rated power.

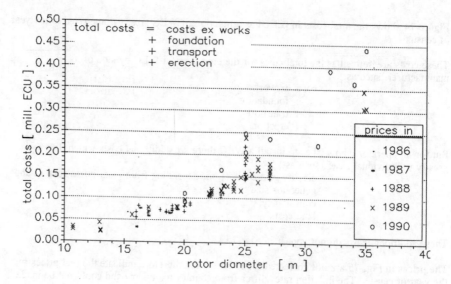

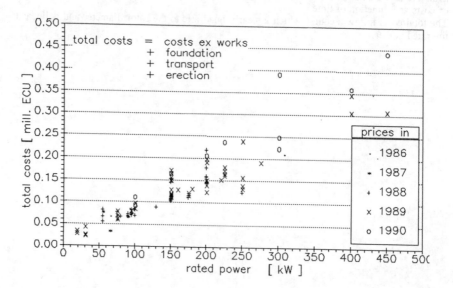

Fig. 75 :   Total costs as a function of WEC size
              (costs do **not** include VAT)

Figure 76 shows the specific total costs in [ECU/m²] and [ECU/kW] for different WEC sizes.

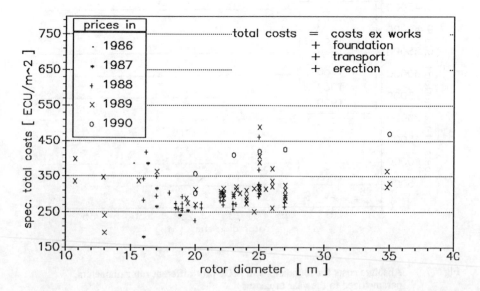

Fig. 76 : Specific total costs in [ECU/m²] and [ECU/kW] as a function of
WEC size, parametrized to the year of census

Figure 77, the last in this section, shows the absolute costs for the foundation as a function of
rotor diameter.

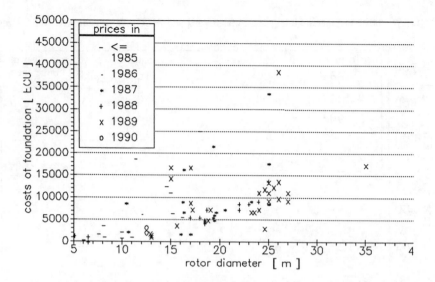

Fig. 77 : Absolute costs for foundations in [ECU] for different rotor diameters, parametrized to the year of census

The cost of foundations depends on conditions at the site, such as stability and load-bearing capacity of the subsoil, and may vary widely. The spread increases with the diameter which means that the security factors differ very much for different foundations. Newer developments regarding foundations, see Pfleiderer (1990), could not yet be considered statistically.

Morthorst;Jensen (1989) give the following survey of the additional costs as percentage of costs ex works (= 100 %) and their development in time:

Table 21 : Relationship between cost components of wind turbines in percentages of the price ex works

vintage	total investment costs	installation and grid connection	foundation	miscellaneous
	%	%	%	%
1980	130.0	18.2	9.0	4.8
1981	131.8	19.4	9.6	6.5
1982	131.2	18.7	8.8	4.3
1983	130.0	17.7	9.1	6.0
1984	125.3	14.4	9.0	3.0
1985	127.6	14.9	9.0	3.1
1986	127.9	15.5	9.1	4.7
1987	129.4	16.5	9.6	3.6
1988	130.6	19.9	8.6	3.9

"The total investment costs of the turbine are approx. 30% above the price of turbine ex works, i.e. approx. 23% of total investment costs are due to foundation, installation/grid connection, and miscellaneous (including consultancy)".

8.2   Annual Energy Output per Unit of Investment

Most important for economic comparison of different energy generation systems are the mean generation costs in monetary unit which depend greatly on annual energy generation per unit of total investment, see equ. (22). This parameter is shown in Figure 78 based on the energy generation of individual WECs in the year 1989 and related to the WEC size. The total investment costs are parametrized to the year of census and they include

Total capital costs

The total incestment or capital costs are the sum of

$$C = C_{EW} + C_F + C_E + C_G + C_M \tag{22}$$

C_{EW} = costs ex works
C_F = foundation costs ,   if not indicated :   9 % of C_{EW}, see Table 21
C_T = transport ,
C_E = erection ,   if C_E and C_T not indicated :   16 % of C_{EW}
C_M = miscellaneous ,   4 % of C_{EW}, see Table 21

Annual energy per investment

$$e_C = \frac{E}{C} = \frac{annual\ energy\ generation}{total\ investment\ costs} \tag{23}$$

Generation costs

$$g = \frac{C}{E} \cdot (K + OM) \tag{24}$$

Annuity factor

$$K = \frac{i(1+i)^n}{(1+i)^n - 1} \tag{25}$$

i   =   annual interest rate on capital

OM =   operational and maintenance costs

n   =   lifetime expectancy

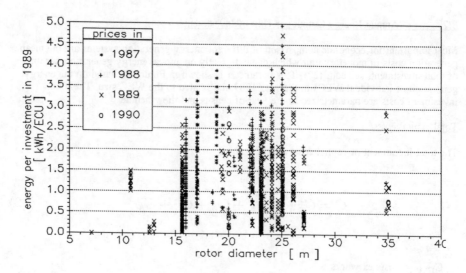

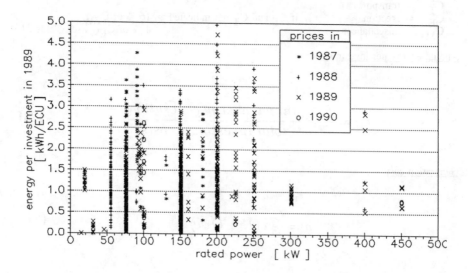

Fig. 78 :   Annual energy in 1989 per unit of investment, parametrized to the year of census, as a function of WEC size

The reciproke values from Fig. 78 represent the only term in equ.(24) which is based on real data, all other quantities such as lifetime expectance, interest rate on capital, annuity factor, O+M cost are derived or approximated.

The real total investment costs by the operators have been available only in very few cases, hence the investment costs of WECs are taken from the sources mentioned at the beginning of this section.

The development of costs in time is also stored continuously, consequently Fig. 77 is parametrized to the year of census. Because some WEC types are no longer manufactured, their total costs are from an earlier stage.

8.3   Generation Costs of Wind Energy

The last link in the diagram of Figure comprises the generation costs which are precisely the mean generation costs of one kWh during the lifetime expectancy. The algorithm after equ. (24) calculates the mean generation costs based on the annuity method. The following assumptions were made to calculate the mean generation costs :

- additional costs for miscellaneous (C_m) from Table 21:	4 % of C_{EW}
- interest rate on capital	5 % per year
- lifetime expectancy	20 years
- operational + maintenance costs	2.5 % of C_{EW} per year

As in Fig. 78 the annual energy generation is based on the year 1989, and the total costs are parametrized to the year of census in Fig. 79.

There is a very wide scatter in mean generation costs, but for the best machines, mean generation costs lower than 0.05 ECU/kWh are possible. This means that wind energy use can be regarded as economically viable when advanced technology and a good local wind potential are combined. The generation costs calculated in this frame consider only the terms mentioned above. A private investor has to consider also local tariffs for the supply of electric energy, tax credits, national support programs for renewable energy sources and more for his individual economic calculation. Of all the renewable electricity generating sources presently available, except hydropower, wind energy use can nowadys be regarded as the most economic, and it becomes even more attractive when social and environmental costs are taken into account.

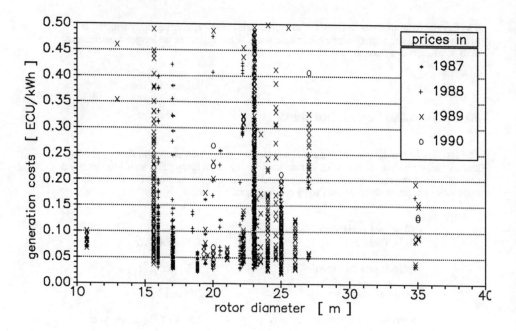

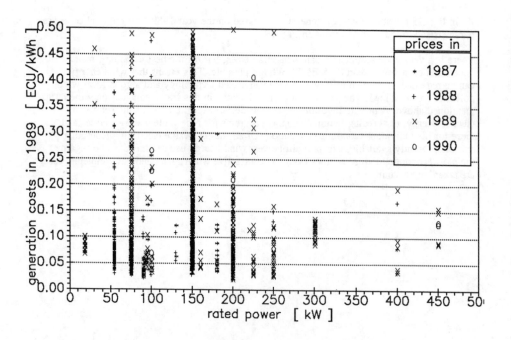

Fig. 79 :   Mean generation costs based on the annual wind energy generation in 1989
parametrized to the year of census, as a function of WEC size.

9   Energy Pay-Back

A question which is often discussed is "How renewable are the "renewables"?", in the
context of this book in particular, the wind energy sources. The question is based on the fact,
that before energy can be generated by a windpower plant, photovoltaic generator, solar
collector, etc, an energy input is neede for manufacturing the energy conversion system. In
general one can say that if there is an economic amortization, even more this is valid for the
energy pay-back time because the energey costs are only a minor part of the total
manufacturing costs of technical products, see Hau (1989).

Different definitions of the energy ratio, the so called "harvest factor", between energy
output and energy input, see equ.(30), are possible. For renewable energy sources the
substitution
method, see Schaefer (1988), which weights the electricity generation of a conversion system
in terms of displaced fossil primary energy, is suitable. The equivalent in displaced fossil
primary energy of the energy output, which considers the total efficiency of the combustion
process, can be approximately obtained by multiplying the electricity generation by a factor
3. Based on the annual energy production of different WECs, see sect. 6.1.2, an answer to
the energy output, that is the saved fossil primary energy, can easily be given by using this
method.
But what about the energety input? The key to the energety input is a knowledge of the
amount and kind of different materials which are necessary for manufacturing a WEC. Here
only a statistical and approximate solution can be given by using statistical factors to
determine the specific energery expenditure in terms of fossil primary energy input for the
fabrication of different materials.
The amount of different materials per WEC follows from sect. 5.4. To calculate the energy
input based on the component weights of section 5.4, some assumptions had to be made:

1.   The generator mass is composed of 50% copper, 50% steel
      generator mass is a linear function of rated power, based on Fig.27

2.   The total rotor mass is composed of 36 % hub (steel) and 64% blades (grp) (statistical
      averages)

3.   The foundation mass (concrete) is equal to the total weight of the WEC

9.1   Specific Primary Energy Input

The overall specific primary energy input for the production of any kind of good is often
named the "hidden energy" or "grey energy" in English literature. The hidden energy
includes all steps of processing, starting with exploitation of raw materials, process energy,
hidden energy of input materials and production equipment. In order to obtain precise and
realistic values, an energety analysis covering the complete process sequence is indispensable.
By monitoring energy consumption on production sites, and with data from the energy and
material statistics of the manufacturing companies, a breakdown of the overall electrical
energy consumption, fuel consumption, as well as non-energetic consumption, e.g. crude
input for the production of plastics, may be calculated. Finally those values in the final
energy sector and non-energetic consumption sector are converted into primary energy
consumption by taking mean values of the overall efficiencies valid for the energy generation
and consumption structure considered.

In the FR Germany these values can be derived from the statistics :

- overall efficiency for generating electrical energy in industry:

$$\eta = 0.4$$

- overall efficiency for supply of fuels:

$$\eta = 0.85$$

- overall efficiency for non-energetic consumption:

$$\eta = 0.8$$

Some previous investigations, see Hagedorn (1989), into the hidden energy of products in the mechanical engineering field have shown that the greatest part of energy consumption is hidden in the preparation of raw materials, whereas additional energy consumption in the subsequent processing steps is usually considerably smaller.

On the basis of past experience and without new investigations, it is possible to obtain a valid approximation of the hidden energy in the mainly used materials by taking mean values for the specific primary energy input, which however include additional charge due to the energy consumption in finishing the product. This charge ranges from 20 to 50 % depending on the scale and complexity of the finishing process.

The approximation of such mean values, which derive from Hau (1989), can be indicated for the dominant materials in WECs as follows:

Table 22 :   Specific primary energy input for technical products consisting mainly of different materials

energy input :

steel	:	**15.5**	**kWh/kg**
copper	:	**25.0**	"
grp	:	**28.0**	"
concrete	:	**0.5**	"

A more significant expression of the specific energety input may be obtained by dividing the cumulative hidden energy by the rated power, see equ.(27). Fig. 80 shows the specific energy input in [MWh$_{Prim. Energy}$/kW$_{Rated}$] of different WECs, based on the procedure explained above.

$$E_{inp} = \sum (W_{mat} \cdot f_{mat}) \tag{26}$$

$$e_{inp} = \frac{E_{inp}}{P} \tag{27}$$

f_{mat} = energy input in terms of primary energy for different materials

m_{mat} = mass of different WEC component materials

E_{inp} = energy input

e_{inp} = specific energy input

P = rated power

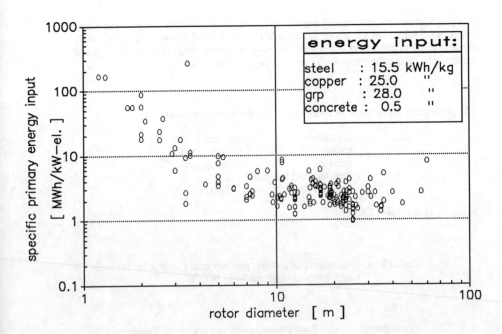

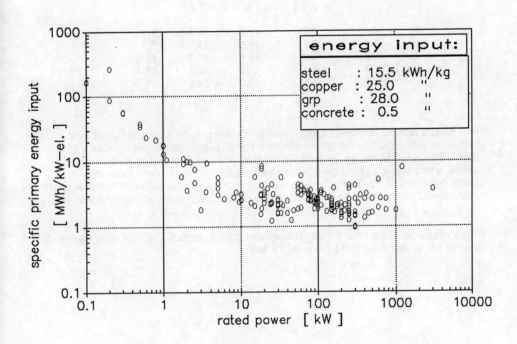

Fig. 80 : Specific cumulative primary energy input of WECs as a function of WEC size

The graph shows an unequivocal degression of specific cumulative energy input towards larger WEC sizes, but a slight increase for WECs from 1000 kW onwards. A possible explanation is that the large-scale machines require the use of different technology and materials to ensure the strength of the much larger consruction.

The percentage distribution of energy input for the main components

- foundation   (concrete)
- tower   (steel/concrete)
- nacelle   (steel/copper)
  incl. hub and sub-components
- rotor blades   (grp)

is shown in the Tabel 23.

Tab. 23 :   Distribution of specific primary energy input for the main components of different sized WECs

rated power class (kW)	foundation (%)	tower incl. hub (%)	nacelle (%)	blades (%)
0 - 10	2.2	43.2	45.2	9.3
10.1 - 75	2.9	49.6	38.2	9.3
75.1 - 300	2.9	48.0	37.7	11.3
300.1 - 1000	3.1	57.9	32.8	17.4
> 1000	3.3	57.2	34.8	10.4

As a general rule the percentage energy input for the tower is the largest, followed by the nacelle (including the hub and subcomponents). For all power classes, the foundation has little influence on the energy input. In Fig. 26 there was a wide scatter in specific nacelle weight, which means that in regard to costs and energy input there is still much scope for design optimization. The tower, as the component with the greatest influence to the energety input, should be investigated more in detail for lattice, tubular or tripod configurations.

The energy input is the basis for the calculation of the harvest facto. and the energy pay-back time which are discussed in the following section.

9.2　　　Harvest Factor and Energy Pay-Back

On the based of the annual wind electricity generation in the year 1988 and the energy input from section 9.1 above, Figure 81 shows the annual harvest factor after equ.(30) for different WECs.

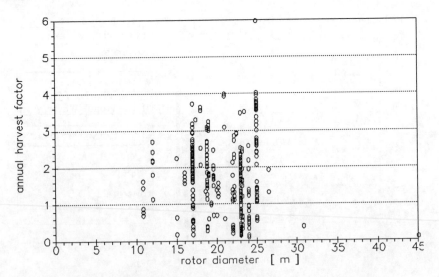

Fig. 81 : Annual harvest factor of different WECs, based on the year 1988

Inversion of the annual harvest factor leads to the energy pay-back, see equ.(31), shown in Fig. 82.

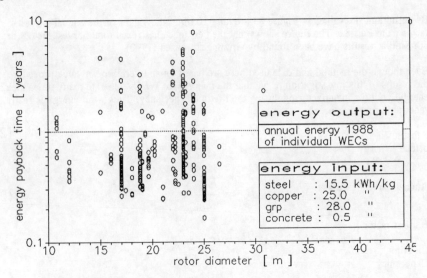

Fig. 82 :　　Energy pay-back time for different WECs based on the energy generation in 1988 and the cumulative hidden energy

Many WECs have an energy pay-back time of less than half a year, a result which makes wind energy the leading renewable energy source as regards regards energy pay-back.

Figure 83 in this section shows the total harvest factor over an assumed lifetime expectancy of 20 years.

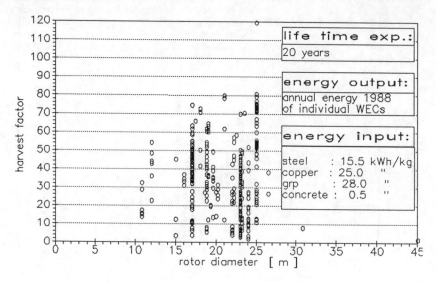

Fig. 83 :   Total harvest factor over the lifetime expectancy for different WECs as a function of the rotor diameter, based on the energy generation in 1988

By definition, the values in Figure 83 are only 20 times the values of the annual harvest factor from Fig. 81. The figure shows that the best machines reach total harvest factors near 80. Similar results have been found by Grum; Schwensen (1990).

Even though the method and data used here are based on an overall mean value procedure, the results of this investigation prove that the majority of WECs generate many times their energy input for manufacturing during the lifetime expectancy. Thus wind energy is really a renewable energy source.

Substituted primary energy

$$E_{subst.} \sim E \cdot 3 \tag{28}$$

Annual harvest factor

$$R_a = \frac{E_{subst.}}{E_{inp}} \tag{29}$$

Harvest factor

Life time Expectancy

$$L$$

$$R = R_a \cdot L \tag{30}$$

Energy Pay-back time

$$T_a = \frac{1}{R_a} \tag{31}$$

10   Summary and Conclusions

Wind energy technology in Europe has been one of the most innovative branches of industry since the beginning of the 80s. A progressive boom in the development, manufacturing and operation of wind turbines can be observed both in the traditionally leading countries such as Denmark and The Netherlands and in the remaining countries of the European Community. Among all the renewable energy resources, the newcomer wind energy now occupies, at the beginning of the 90s, a favourable position as regards both research and exploitation of the enormous wind energy potential of Europe.

This book surveys the important start phase of wind energy technology in Europe between 1986 and 1989 by quantifying the significant facts and trends.

There are now more than 3200 units throughout Europe with a total installed capacity of nearly 390 MW stored in the EUROWIN database. The monthly reporting quote has increased continuously and at the end of the evaluation period of this book in December 1989, nearly 2000 units, more than twice the total in January 1986, reported monthly data to the database. The significant design parameters are plotted in master tables and the most important ones such as power curves, design wind speeds, component weights, hub height, rated power, etc. are plotted graphically.

Most important are the evaluations of operational data. The cumulative energy generation shows a typical cyclic pattern in the yearly profile, with characteristic winter maxima and summer lulls. Very few wind measurement data have been available, therefore a statistical investigation of how the wind potential affected the energy generation was not carried out. This will be object of future investigations when more wind measurement data become available.
The specific energy, expressed as annual energy generation of a WEC divided by its rotor swept area, is a good indication to compare different WEC sizes or technologies. The best machines had specific energies between 1000 and nearly 1600 [kWh/m² a] and they are represented by nearly 2 % of all WECs evaluated. The majority of machines have annual specific energies between 500 and 700 [kWh/m² a].
The influence of the hub height and the ratio of hub height to rotor diameter has been quantified. Up to a hub height of 30 m an increase in specific energy is observed, whereas beyond 30 m hub height the evaluated WECs had lower values.
The impact of total operational time on the energy generation shows that the new machines, erected at the end of the 80's, are significantly better than machines erected at the beginning of the 80's.
A comparison of the actual energy generation of an operating WEC with the prospective annual energy generation, mostly indicated by the manufacturer, shows that in all the years evaluated, the average conformity of all reported WECs was less than 100 %.

As for the energy generation, the corresponding capacity factor and the operational time were calculated both for individual WECs and for the average of all machines.

Failure analysis of eight main components was carried out. The control system and the yaw system were the most faulty components. The influence of failure frequency on the corresponding downtime could not be investigated since too few data were available. This influence will be investigated from 1990 on, when more downtime data have been reported.

In the chapter on economics, the absolute and specific costs, the annual energy per unit investment and the mean generation costs are quantified. For the best a machines at mean generation cost approaching 0.05 ECU/kWh is possible. This means that wind energy can be a viable alternative to coal-fired or nuclear power plants at sites where advanced technology and a good local wind energy potential are combined.

In the last chapter the specific primary energy input and the energy pay-back time of WECs are quantified on the basis of the operational data and the mass distribution of WEC components. Total harvest factors of 80 are possible. The majority of WECs generate many times their energy input for manufacturing.

On the basis of the results presented, the following main fields should be investigated in more detail as regards energy generation or failure analysis :

- wind measurement data

- windparks

- component failures such as
  - gearbox
  - hub
  - control system

Statistical investigation of wind energy development is a necessary accompaniment to the actual operation of modern WECs. We hope this book will provide the reader with all the information needed to form an opinion and complete his knowledge about wind energy use.

11 References

Brabandt, B. ; Möller, U. ;
"Der Erntefaktor- Ein Schlüssel zum wirkungsvollen Energieeinsatz?"
Brown, Boveri & Cie AG, Kallstadter Str. 1, D- 6800 Mannheim

Bussel, G. van ; 1989
"Wind Turbine monopod Foundation developped"
Wind Stats, autumn 1989, p.1-2

FDV/Verein Dänischer Hersteller von Windkraftanlagwen/FDV ; 1990
"Die Windkraftanlagen werden größer und der Erzeugerpreis hat sich seit dem Jahre 1980 auf
ein Drittel reduziert"
Wind Kraft Journal 2/90

Follings, F.J. ; 1989
"Economic optimization of wind power plants"
EWEC 1989, Glasgow, p. 983
Peter Peregrinus Ltd,, U.K.

Frandsen, S. ; Madsen, P.H. ; 1989
"Optimization of Power Output from Wind Turbines"
Riso National Laboratory Denmark
EWEC 1989, Glasgow, p. 1006
Peter Peregrinus Ltd., U.K.

Grum-Schwensen, E. ; 1990
"The Real Cost of Wind Turbine Construction"
Wind Stats, spring 1990, vol.3, No.2, p.1-2

Harrison, R. ; Jenkins, G. ; Taylor, R.J. ; 1989
"Cost modelling of horizontal axis wind turbines - results and conclusions"
Sunderland Polytechnic U.K., ETSU Harwell U.K.
EWEC 1989, Glasgow, p. 988
Peter Peregrinus Ltd., U.K.

Hau, E. ; 1984
"What is the most economical size of a large wind power < plant?"
proceedings of EWEC 1984, Hamburg, p. 820

Hau, E. ; 1989
"Windkraftanlagen"
Springer Verlag

Hagedorn, G. ; 1989
"Hidden energy in solar cells and photovoltqaic power stations"
Proc. 9th European PV Solar Energy Conference

Heier, S. ; 1989
"Informationspaket Nutzung der Windenergie"
BINE, Fachinformationszentrum Karlsruhe, Büro Bonn, Ahrstr. 64, D-5300 Bonn

Hensing, P.C. ; Steen, A. van der ; in (Nacfaire, H.;CEC;1988)
"Development and production of a cost-effective 1 MW windturbine"

Honnef, H. ; 1988
"Windkraftwerke"
facsimile DGW

Informationssekretariatet for Vedvarende Energi ; 1986 ; 1988 ; 1989
"Vindmolle Oversigt Juni 1986, Juni 1988, Sept. 1989"
Teknologisk Institut
Gregersensvej
2630 Tastrup

Interressenverband Windkraft e.V. ; 1989
"Marktübersicht 1989"
Fahlbachweg 94
4532 Mettingen

Jarass, L. ; 1981
"Strom aus Wind, Integration einer regenerativen Energiequelle"
Springer-Verlag

Jarass, T.F. ; 1987
"Wind Eneregy 1987, Wind Turbine Shipments and Applications"
Wind Data Center, Stadia Inc., P.O.Box 442, Great Falls, Virginia 22066, U.S.A.

Klein, H.P. ; 1986
"Über die spezifische Jahresenergie in Abhängigkeit bestimmter äußerer Randbedingungen"
Wind Kraft Journal, 2/86

Klein, H.P. ; 1987
"Einfluss von Kennlinie und Entwurfsparametern auf die spezifische Jahresenergie
stromproduzierender Windkraftanlagen"
Tagung WINDENERGIE, Universität Oldenburg, 27/28 März 1987

Landwirtschaftskammer Schleswig-Holstein ; 1989
"Windenergie II-Praxisergebnisse 1989"
Kiel

Leuven, J. van ; 1984
"The economic feasibility of small-scale wind energy conversion systems"
proceedings of EWEC 1984, p.814

Milborrow, D.J. ; 1989
"Will Multi-Megawatt Wind Turbines ever be economic ?"
Central Electricity Generating Board, U.K.
EWEC 1989, Glasgow, p. 1017
Peter Peregrinus Ltd., U.K.

Milborrow, D. ; 1990
"Wind energy during the 1980s"
windirections spring 1990

Molly, J.P. ; 1989
"Maximum economic Size of Wind Energy Converters"
WISA-Energiesysteme GmbH, FRG
EWEC 1989, Glasgow, p. 1013
Peter Peregrinus Ltd., U.K.

Molly, J.P. ; 1990
"Windenergie in Theorie und Praxis"
C.F.Müller Verlag

Mortensen, K. ; Skamris, C. ; in (Nacfaire, H.;CEC;1988)
"The Masnedo Wind Farm"

Morthorst, P.E. ; Jensen, P.H. ; 1989
"Economics of wind turbines"
Riso National Laboratory

Nacfaire, H. ; CEC ; 1988
"Grid Connected Wind Turbines",
Elsevier Applied Science

Nacfaire, H.N. ; Diamantaras, K. ; 1989
"The European Community Demonstration Programme for Wind Energy and Community
Energy Policy"
EWEC 1989, Glasgow, p. 1
Peter Peregrinus Ltd., U.K.

Nordvestjysk Folkecenter ;
"Oversigt over Danske Vindmoller"
Kammershardsvej 16
DK-7760 HUROP Thy

Pfleiderer Consulting ; 1990
"Cement Towers for Turbines"
Wind Stats, autumn 1990, p.2-3

Riegler, G. ; 1983
"Windturbinen im Höhenkraftwerk, Grundlagen über den Energieentzug aus einer
Windströmung"
dbv-Verlag Graz

Schaefer, H. ; 1988
"Erntefaktoren von Kraftwerken"
Energiewirtschaftliche Tagesfragen, H.10

Schmid, J. ; 1984
"Important Results of the European Wind Energy Programme"
EWEC Hamburg, p.910-916

Schmid, J. ; 1986
"European Wind Energy Technology, State of the art of wind energy converters in the
European community"
D.Reidel Publishing Company

Schmid, J. ; Klein, H.P. ; 1987
Presentation of the European Wind Turbine Data Base EUROWIN
ISES Solar World Congress, Hamburg

Schmid, J. ; Klein, H.P. ; 1988
"Databank on existing Windturbines and Climates in the European Community"
EWEC 1988 Herning

Schmid, J. ; Klein, H.P. ; 1989
"EUROWIN the European Wind Turbine Data Base goes 'on line' "
windirections, spring 1989

Schmid, J. ; Klein, H.P. ; 1990a
"Energy and costs, the most important parameters to the use of wind energy"
windirections, winter 1989/1990

Schmid, J. ; Klein, H.P. ; 1990b
"Review of European Wind Energy Technology"
ECWEC 1990, Madrid

Schmid, J. ; Klein, H.P. ; Godard, H. ; 1988a
"Energie und Kosten, die wichtigsten Parameter um den Stellenwert der Windenergie"
Wind Kraft Journal, 4/88, S.178-181

Schmid, J. ; Klein, H.P. ; Godard, G. ; 1988b
"Eurpean Wind Turbine Data Base EUROWIN"
Euroforum Saarbrücken, 24.10.- 29.10.1988

Schmid, J. ; Klein, H.P. ; Hagedorn, G. 1991
"News from EROWIN, 'How renewable is wind energy ?' "
WINDirections winter 1990/91

Schönball, W. ; 1989
"Windenergie Jahrbuch 1988/89"
Verlag C.F.Müller GmbH, Karlsruhe

Seifritz, W. 1978
"Sanfte Energietechnologie-Hoffnung oder Utopie?"
Band 92, Verlag Kark Thiemig, Munich

141

Seifritz, W. ; 1980
"Zur Dynamik von Substitutionsprozessen"
Eidg. Institut für Reaktorforschung
CH-5303 Würenlingen

Warne, D.F. ; 1983
"Wind Power Equipment"
E. & F.N. SPON Ltd., London/New York

APPENDIX

A   Description of the EUROWIN Database

A.1   Data Processing

This appendix sets out the data processing procedure between the origin input data and data output, see Fig. 84. The center and base for all procedures is the EUROWIN database. In the periphery different programs are run for data input, plausibility check and calculations. The last link is the graphic output of special data inquiries or derived quantities.

The origin data coming from diskette or questionnaire must be converted to the database structure by converting programs or data input masks. The indications from the origin data are often equivocal. A plausibility check according to different criteria avoids adding faulty input data to the database. For example, if operational data have an indication about monthly energy production but "0" for operational time, the plausibility check has to distinguish between "zero" and "no indication". Operators sometimes give no monthly operational data in a certain month (absence, holidays, etc.) and in the following months they indicate the operational data for a two-month period. In that case the operational data over- or understep certain limits set by the plausibility check which makes an individual decision necessary. Wrong indications within the set limits cannot be identified by this procedure.
The checked data are added and distributed to one of the several files in the data base, see Appendix A.2. Based on these monthly data, several programs analyse the database according to different criteria. The results of these calculations are also stored in the database in special files which contain accumulative data, averages or other derived quantities. The results can be presented by listing or graphic format.

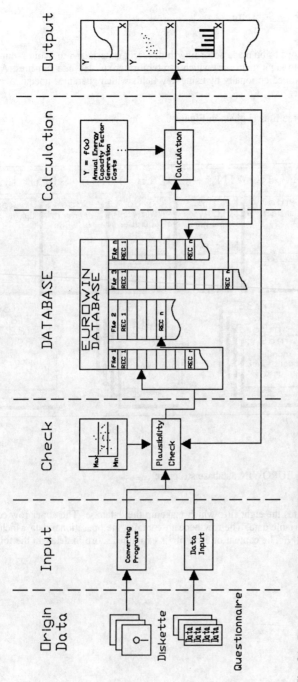

Fig. 84 : Data processing within the EUROWIN database

A.2   Data Base Structure

To permit evaluations to be made according to different criteria and to avoid redundancy, the database is split into eight main files which can be further splitted into sub-files. All files can be combined and joined optionally by using key-fields which guarantee unequivocal access between each other.

An overview of these files is given in Figure 85.

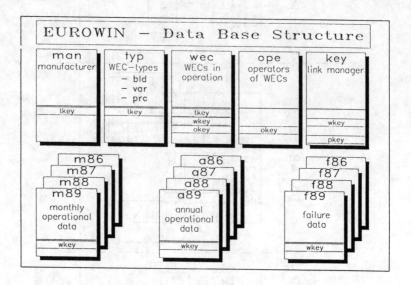

Fig. 85 :   EUROWIN database structure

The figure illustrates the eight files which make up the database. The upper row contains files with fixed or invariable data. The row beneath contains the operational data which increase monthly or annually. The content of each file is explained more in detail in the following pages.

name of file	content

man   identification and invariable manufacturer data such as
- address
- telephone, telex, telefax

typ   technical data of a WEC-type such as
- rated power, rotor diameter, hub height
- control principle, overspeed protection
- transmission data, brake data, generator data, tower data
- component materials, component weights
- design wind speeds, power curve, calculated annual energy, etc.

sub files:
- **bld** (blades)   - blade geometry
- **var** (variation)   - variation of type (tower type, tower height, etc.)
- **prc** (prices)   - prices obtained from market analysis

WEC   fixed data of an operating WEC such as
- WEC-type identification data
- location of WEC
- rated power, rotor diameter, hub height
- mean site wind speed, Weibull parameters
- expected annual energy
- installation costs

ope   fixed data of operator (one operator can have several WECs at different locations) such as
- name, address, contact person, etc.

key   this file manages the unequivocal access between different files and records for determined keys such as
- **tkey** (type key) containing
  - country key
  - manufacturer key
  - type key
  - version key (type version)
  - variation key (type variation)
  - modification key (type modification)

- **wkey** (WEC key) containing
  - origin national key
  - current number of WEC
  - current number of windpark

- **okey** (operator key) containing
  - current number of operator

- **nkey** (national key)   controls the access between origin national keys and EUROWIN keys

month   monthly operational data of operating WECs from
- origin data such as
    - energy production,
    - reported component failures
    - operational time

- derived quantities such as
    - specific energy
    - capacity factor

annual   cumulative annual operational data obtained from monthly data such as
- absolute and specific annual energy
- average annual capacity factor
- average operational time

fail   failure analysis data such as
- absolute and relative failure distribution of various WEC components
- failure distribution as a function of operational time
- classification of failures to cause and effect

A.3   Data Base Operation

INFO   RETRIEVE   INPUT   AUTO-INPUT   DIAGRAM   DOCUM   EVAL   LIST   17:23:05

EUROPEAN WIND-ENERGY DATABASE   c/o FhG-ISE D-7800 Freiburg

Fig. 86 : Main menu of EUROWIN database

EUROWIN is a relational database written in INGRES for UNIX and in dBase IV for MS-DOS. The operation of the database is menu led, see Fig. 89. The first line indicates the eight pull-down menus which an operator can choose.

INFO   :   gives a quick survey of the status of the database, see Table 1
                        sub-menus :   - contents
                                      - update
                                      - environment
                                      - date
                                      - end

RETRIEVE :   contains the fixed data, see Fig. 87
                        sub-menus:
                        - MAN
                        - TYP
                        - OPE
                        - WECs
                        - KEY

INPUT :   manual input of new data, the same as RETRIEVE, plus
                        - MONTHLY

AUTO-INPUT : input by floppy disc
sub-menus :
- NL new WECs
- NL monthly data
- DK new WECs
- DK monthly data

DIAGRAM   : administration of graphics

DOCUM   : documentation about the database, such as
- file structures
- country keys and currencies

EVAL   : evaluation programs, such as
- cumulative energy
- capacity factor
- generation costs
- etc.

LIST   : prepared listings from the database

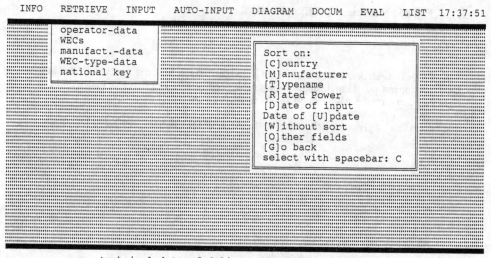

```
INFO    RETRIEVE    INPUT    AUTO-INPUT    DIAGRAM    DOCUM    EVAL    LIST   17:37:51
```

```
operator-data
WECs
manufact.-data
WEC-type-data
national key
```

```
Sort on:
[C]ountry
[M]anufacturer
[T]ypename
[R]ated Power
[D]ate of input
Date of [U]pdate
[W]ithout sort
[O]ther fields
[G]o back
select with spacebar: C
```

```
technical data of delieverable windturbine-types
```

Fig. 87 : Example of pull-down menu RETRIEVE

Figures 88 - 90 are examples of data sheets obtained by inquiry from the EUROWIN database.

```
INFO   RETRIEVE   INPUT   AUTO-INPUT   DIAGRAM   DOCUM   EVAL   LIST   17:31:06
```

```
Country:  Denmark              *** TYP ***   151/733    Indate:   07.07.88
Manufact: Bonus Energy A/S (BONUS)                      Update:   30.05.90 (bah)
Typename: Bonus 150 kW   Version: 1  Number: 01         TKEY:    02 020 212 1 01
Status:   1 at: 10.12.89 ──────────────────────────────────────────────────────
P-rated:   150,00 kW (PC  5) Diam:  23,0 m (DC  9)     Rotor aerea:   415,0 m²
P-spec:    361,4 W/m²  Max. pow. C: 0,000   P-peak:  165,0 kW    Admission: 1
──────────────────────────────────────────────────────────────────────────────
Rotor and blades:
Number of rotors: 1  Rotor-typ:  1  Number of blades:  3   Lambda:       0,0
Profile: NACA 63-200    Manuf. of blades: LM              Blade mater.: 16
Tip angle: 1  Tip 1: 16,00 Tip 2:  0,00   Rot.orient.: 1  Yaw system:    1
Rot.speed: 2  Speed low:  30 Speed high:  40 rpm          Hub-typ:       1
Power control: 1  Overspeed protection: 264      Power curve available
──────────────────────────────────────────────────────────────────────────────
Transmission :
Trans. typ:  4 Ratio: 1 1:24,8                    Brake typ: 3 position:  1
──────────────────────────────────────────────────────────────────────────────
Generator:                           P-rated    Speed   Cos-phi Poles
Number of gen.: 3 Typ:  18   Small gen.  30,00 kW  750 rpm   0,00    6
Frequ: 50 Hz   Gen.-use:   Main gen.   150,00 kW 1000 rpm   0,93    6
```

```
<↑>/<↓>,<PgUp>/<PgDn>  more typedata=F4  manufacturer=F5  print=F6  term.=Esc
```

Fig. 88 :   Example of data sheet obtained from the EUROWIN database

```
INFO   RETRIEVE   INPUT   AUTO-INPUT   DIAGRAM   DOCUM   EVAL   LIST   17:37:15
```

```
Country:  Belgium             *** TYP ***   6/733      Indate:   05.06.88
Manufact: HMZ Belgium N.V. (HMZ)                        Update:   20.04.90 (hgo)
Typename: Windmaster 300  Version: 2  Number: 01        TKEY:    01 001 007 2 01
Status:   1 at: 25.09.89 ──────────────────────────────────────────────────────
P-rated:   300,00 kW (PC  6) Diam:  25,0 m (DC  9)     Rotor aerea:   480,8 m²
P-spec:    624,0 W/m²  Max. pow. C: 0,000   P-peak:  300,0 kW    Admission: 1
──────────────────────────────────────────────────────────────────────────────
Data of type-variations:   1 of  2   Input: 05.06.88 Update: 10.04.90 (hgo)
Tower     typ: 3 material: 1 height:   21,8 m hub-height:  23,0 m
Weights tower:  9000kg nacelle:   6200kg rotor:   3960kg blade:    820kg
      generator:  1400kg    head: 10160kg total:  19160kg
──────────────────────────────────────────────────────────────────────────────
Costs .BFr .25.09.89│.........................................................
tower ...... 1207985│.........................................................
machine ...  9161021│.........................................................
transport ..       0│.........................................................
foundation .       0│.........................................................
erection ...       0│.........................................................
total .. 3. 10369006│.........................................................
service/a ..       0│.........................................................
```

```
previous variation=F2   next variation=F3   more typedata=F4
```

Fig. 89 :   Example of data sheet obtained from the EUROWIN database

```
INFO   RETRIEVE   INPUT   AUTO-INPUT   DIAGRAM   DOCUM   EVAL   LIST  17:35:00
═══════════════════════════════════════════════════════════════════════════════
Country:  FR Germany    *** TYP ***   667/733      Indate:   15.03.90
Manufact: Enercon (ENERC)                           Update:   01.06.90 (hgo)
Typename: Enercon 32/300  Version: 1  Number: 01    TKEY:     05 006 006 1 01
Status:   1  at: 01.09.89 ──────────────────────────────────────────────────────
P-rated:  300,00 kW (PC 6) Diam:  32,0 m (DC 10)    Rotor aerea:  804,0 m²
P-spec:   373,1 W/m²  Max. pow. C: 0,000   P-peak:  300,0 kW   Admission: 1

Power curve:   (measured by:  1)

  V (m/sec)   0,5     1,5     2,5     3,5     4,5     5,5     6,5     7,5
  P (kW)      0,00    0,00    0,00    8,60   18,40   34,90   55,20   84,13

  V (m/sec)   8,5     9,5    10,5    11,5    12,5    13,5    14,5    15,5
  P (kW)    122,22  169,84  225,40  300,00  300,00  300,00  300,00  300,00

  V (m/sec)  16,5    17,5    18,5    19,5    20,5    21,5    22,5    23,5
  P (kW)    300,00  300,00  300,00  300,00  300,00  300,00  300,00  300,00

  V (m/sec)  24,5    25,5    26,5    27,5    28,5    29,5
  P (kW)    300,00    0,00    0,00    0,00    0,00    0,00
```

next type=F2   previous type=F3   more typedata=F4   terminate=Esc

fig. 90 : example of data sheet, obtained from EUROWIN data base

B   List of WEC Types and WEC Manufacturers

Type name	manufact. short name	manufacturer name	country
(55/16)	POLENK	Polenko B.V.	NL
04-03	VENTIS	Ventis Energietechnik	FRG
1 KW Anlage	HUELL	Huellmann-Anlagenbau KG	FRG
1.1 kW	PROV	Proven Eng. Prod. Ltd.	UK
10 J 18	DEJONG	Konstruktiebedrijf De Jong	NL
10 WPX	FDO	Stork FDO WES B.V.	NL
10/30	BORN	Born F., Holzleimbau	FRG
10/7.5	WINFOS	Windfos A/S	DK
100	AMPAIR	Ampair	UK
1000.9	PROENE	Proenerga, S.A.	E
100KW	BALPOW	Baltic Power A/S	DK
100W	AMPAIR	Ampair	UK
10kW8	HMZ	HMZ Belgium N.V.	B
11/15	WYNFAN	Wynfang	NL
11/17.5	WIPAQU	Paques BV	NL
11/7.4	ALTERN	Alternegy Aero Star	DK
11m-22kW	BOUMA	Bouma Windenergie B.V.	NL
11m-22kW	BOUMA	Bouma Windenergie B.V.	NL
11m-30kW	BOUMA	Bouma Windenergie B.V.	NL
12.5 WPX	FDO	Stork FDO WES B.V.	NL
12/30	ECOTEC	Ecotecnia s. coop.	E
125/50 G 30	KROG	Krogmann Maschinenbau	FRG
15-75	ALTERN	Alternegy Aero Star	DK
15-90	ALTERN	Alternegy Aero Star	DK
15/8.8	WINFOS	Windfos A/S	DK
150-7	AERWAT	Aerowatt International	F
150.7	PROENE	Proenerga, S.A.	E
150/19.6	ALTERN	Alternegy Aero Star	DK
15K	KURI	Kuriant-Maskinfabrik	DK
16m-75Kw	BOUMA	Bouma Windenergie B.V.	NL
1800 Series	MARLEC	Marlec Engeneering Co. Ltd.	UK
1800 W	GEMZ	GEMZ	E
19/110	DANWIN	Danwin A/S	DK
19/120	DANWIN	Danwin A/S	DK
1kW	PRIMA	Primary Energy Co-op. Society	IRL
20 WPX	FDO	Stork FDO WES B.V.	NL
20-100	VENTIS	Ventis Energietechnik	FRG
20/150	ECOTEC	ecotecnia s. coop.	E
200 kW	HAWKER	Hawker Siddeley Power Eng. Ltd	UK
200/23.3	ALTERN	Alternegy Aero Star	DK
2000 W	GEMZ	GEMZ	E
200kW	FLEMA	FLEMA A/S	DK
200kW/415m^2	WIWOR	Wind World A/S	DK
20KW12	HMZ	HMZ Belgium N.V.	B
23/140	DANWIN	Danwin A/S	DK
24 kW	HMZ	HMZ Belgium N.V.	B
24.5m/250kW	BOUMA	Bouma Windenergie B.V.	NL
25 KW aerodyn	AERODY	Aerodyn Energiesysteme GmbH	FRG
25kW	NEI	NEI International Res.& Devel.	UK

B List of WEC Types and WEC Manufacturers (contd.)

Type name	manufact. short name	manufacturer name	country
275/25.3	ALTERN	Alternegy Aero Star	DK
3 KW	BALTIC	Baltic Korn A/S	DK
3.4.5	CAST	CAST	I
30/10.8	ALTERN	Alternegy Aero Star	DK
30/11.2	WINFOS	Windfos A/S	DK
35 SI 500	NEWINC	Newinco BV	NL
400-7	AERWAT	Aerowatt International	F
45 kW	MULTI	Multimetaal Constructie B.V.	NL
45/12.5	ALTERN	Alternegy Aero Star	DK
48211	CEMEL	C:E:M:E:L: 83 s.n.c.	I
5 WPX	FDO	Stork FDO WES B.V.	NL
5/5.5	BORN	Born F., Holzleimbau	FRG
500 W	GEMZ	GEMZ	E
500 W	TORNA	Tornado Wind Generators, Ltd.	UK
55/11	FAST	Fasterholt Maskinfabrik A/S	DK
56-100	AWPP	AWP Plantas Eolicas	E
6.3 WPX	FDO	Stork FDO WES B.V.	NL
6.6/11	BORN	Born F., Holzleimbau	FRG
6/10	ISEA	I.S.E.A. S.R.L.	I
60-7	AERWAT	Aerowatt International	F
60.7	PROENE	Proenerga, S.A.	E
60kW	NEI	NEI International Res.& Devel.	UK
65 KW Ruhrtal	RUHRTA	Ruhrtal-Elektrizitaetsgesellschaft	FRG
65/16	ALTERN	Alternegy Aero Star	DK
75 kW	HANING	Handel- og Ingeniersfirma	DK
75/15	FAST	Fasterholt Maskinfabrik A/S	DK
75/17	WINFOS	Windfos A/S	DK
750kW	ELKRAF	Elkraft AMBA	DK
7M	AEOLTD	Aeolus Ltd.	UK
8 WPX	FDO	Stork FDO WES B.V.	NL
8/15	BORN	Born F., Holzleimbau	FRG
80kW	DWTNAK	Danish Wind Turbines Nakskov	DK
9/15 kW	MMEW	MMEW Marketing Ltd.	UK
910 Series	MARLEC	Marlec Engeneering Co. Ltd.	UK
99kW/19.6m	SOEG	Soegrens Maskinfabrik	DK
AC55	AERCAN	Aerogeneradores Canarios	E
Adler 25/100kW	KOEST	Koester Maschinenfabrik/Adler Windtechn.	FRG
Adler 25/165kW	KOEST	Koester Maschinenfabrik/Adler Windtechn.	FRG
Aeolus 11	AERODY	Aerodyn Energiesysteme GmbH	FRG
Aeolus 12/220V	AERODY	Aerodyn Energiesysteme GmbH	FRG
Aeolus 12/380V	AERODY	Aerodyn Energiesysteme GmbH	FRG
Aeolus WTS-75	MBB	Messerschmitt-Boelkow-Blohm GmbH	FRG
AERO 14PI50/65	NEWINC	Newinco BV	NL
AERO 17PI85/100	NEWINC	Newinco BV	NL
Aero Tech 75	ADWIP	Advanced Windpower Products A/S	DK
AERO23PI200/250	NEWINC	Newinco BV	NL
AERO34PI500	NEWINC	Newinco BV	NL

B   List of WEC Types and WEC Manufacturers (contd.)

Type name	manufact. short name	manufacturer name	country
Aeroman 11/20	MAN	MAN Technologie GmbH	FRG
Aeroman 12.5/30	MAN	MAN Technologie GmbH	FRG
Aeroman 12.5/33	MAN	MAN Technologie GmbH	FRG
Aeroman 12.5/40	MAN	MAN Technologie GmbH	FRG
Aeroman 12/20	MAN	MAN Technologie GmbH	FRG
Aeroman14,8/33i	MAN	MAN Technologie GmbH	FRG
Aeroman14.8/33n	MAN	MAN Technologie GmbH	FRG
AEROT 23PI/250	NEWINC	Newinco BV	NL
AEROTE14PI50/65	NEWINC	Newinco BV	NL
Aerotech 10	TCR	TCR Techn. Combinatie Rhenen	NL
AEROTECH 10PI20	NEWINC	Newinco BV	NL
AEROTECH 10PS20	NEWINC	Newinco BV	NL
Aerotech 14	TCR	TCR Techn. Combinatie Rhenen	NL
AEROTECH 14PS50	NEWINC	Newinco BV	NL
Aerotech 17	TCR	TCR Techn. Combinatie Rhenen	NL
AEROTECH 17PS85	NEWINC	Newinco BV	NL
Aerotech 23	TCR	TCR Techn. Combinatie Rhenen	NL
AEROTECH34PI500	NEWINC	Newinco BV	NL
Aiolos 100 kW	HELLAI	Hellenic Aerospace Industry ltd	GR
Aiolos 55 kW	HELLAI	Hellenic Aerospace Industry ltd	GR
AIT 02 - Medit	AERIT	Aeritalia S.A.I. S.p.A.	I
AIT 02 - Medit	AERIT	Aeritalia S.A.I. S.p.A.	I
AIT 03	AERIT	Aeritalia S.A.I. S.p.A.	I
AWP 200/23	ADWIP	Advanced Windpower Products A/S	DK
AWP 90/18	VISYS	Vind-Syssel A/S	DK
AWP 90/18	ADWIP	Advanced Windpower Products A/S	DK
B 25	SUEWIN	Suedwind Windkraftanlagen	FRG
B 30	SUEWIN	Suedwind Windkraftanlagen	FRG
BBW 20KDG	WISO	Windsol Ltd.	GR
BBW 30KDG	WISO	Windsol Ltd.	GR
BON 100/19.4	BONUS	Bonus Energy A/S	DK
BON 100/20	BONUS	Bonus Energy A/S	DK
BON 100/23	BONUS	Bonus Energy A/S	DK
BON 150/23	BONUS	Bonus Energy A/S	DK
BON 190/19.4	BONUS	Bonus Energy A/S	DK
BON 22/10	BONUS	Bonus Energy A/S	DK
BON 22/12	BONUS	Bonus Energy A/S	DK
BON 30/10	BONUS	Bonus Energy A/S	DK
BON 30/11	BONUS	Bonus Energy A/S	DK
BON 300/35	BONUS	Bonus Energy A/S	DK
BON 45/15	BONUS	Bonus Energy A/S	DK
BON 450/35	BONUS	Bonus Energy A/S	DK
BON 55/11	BONUS	Bonus Energy A/S	DK
BON 55/12	BONUS	Bonus Energy A/S	DK
BON 55/15	BONUS	Bonus Energy A/S	DK
BON 55/16.5	BONUS	Bonus Energy A/S	DK
BON 55/18	BONUS	Bonus Energy A/S	DK
BON 95/19.4	BONUS	Bonus Energy A/S	DK
Bost 18/10.9	BOST	Bosted Moellen ApS	DK

B   List of WEC Types and WEC Manufacturers (contd.)

Type name	manufact. short name	manufacturer name	country
Bost 18/10.9	BOST	Bosted Moellen ApS	DK
Bosted Moellen	BOST	Bosted Moellen ApS	DK
BOUMA 160/20	BOUMA	Bouma Windenergie B.V.	NL
BW 2KB	WISO	Windsol Ltd.	GR
BW 10KC	WISO	Windsol Ltd.	GR
BW 10KCG	WISO	Windsol Ltd.	GR
BW 1KA	WISO	Windsol Ltd.	GR
BW 3KB	WISO	Windsol Ltd.	GR
BW 5KC	WISO	Windsol Ltd.	GR
BW08	BRUEM	Bruemmer Windkraftanlagen KG	FRG
BW120	BRUEM	Bruemmer Windkraftanlagen KG	FRG
BW160	BRUEM	Bruemmer Windkraftanlagen KG	FRG
BW81	BRUEM	Bruemmer Windkraftanlagen KG	FRG
BWC 1000	BERGEY	Bergey Windpower	USA
BWC 1500	BERGEY	Bergey Windpower	USA
BWC EXCEL	BERGEY	Bergey Windpower	USA
C. 100-12	HARB	A. Harbarth	FRG
Camperduin	LAWEY	Lagerwey Windturbine	NL
CCFI	COMHAR	Comhar Chumann Bun Fhutnnimh T	IRL
CE.8000	CRESS	Cresswell Engineering	UK
Co-Dan 200 kW	CO-DAN	Co-Dan	DK
D 110	DWP	Danish Windpower DWP	DK
D 140	DWP	Danish Windpower DWP	DK
D 150	DWP	Danish Windpower DWP	DK
D 175/23.6	DWP	Danish Windpower DWP	DK
D-200	DENCON	Dencon Holding A/S	DK
D-250	DENCON	Dencon Holding A/S	DK
D-75	DENCON	Dencon Holding A/S	DK
Danwin 19/140	DANWIN	Danwin A/S	DK
Danwin 23/180	DANWIN	Danwin A/S	DK
Danwin 24/150	DANWIN	Danwin A/S	DK
Danwin 25/250	DANWIN	Danwin A/S	DK
Danwin 27/225	DANWIN	Danwin A/S	DK
Darrieus	HKW	HK Warmtemolen	NL
Darrieus DZ-12	DORN	Dornier System GmbH	FRG
DVI 15/3	GIRO	Giromill	NL
DWT 30/265	DWT	Danish Wind Technology A/S	DK
DWT 9.6/18	DWT	Danish Wind Technology A/S	DK
E 1220	SUEWIN	Suedwind Windkraftanlagen	FRG
E 1225	SUEWIN	Suedwind Windkraftanlagen	FRG
E 305	SUEWIN	Suedwind Windkraftanlagen	FRG
E 710	SUEWIN	Suedwind Windkraftanlagen	FRG
Elektromat 1 kW	WIKZ	Windkraft-Zentrale	FRG

B   List of WEC Types and WEC Manufacturers (contd.)

Type name	manufact. short name	manufacturer name	country
ENERCON 16/55	ENERC	Enercon Ges. f. Energieanlagen mbH&Co	FRG
ENERCON 17/80	ENERC	Enercon Ges. f. Energieanlagen mbH&Co	FRG
ENERCON 32/280	ENERC	Enercon Ges. f. Energieanlagen mbH&Co	FRG
ENERCON 32/300	ENERC	Enercon Ges. f. Energieanlagen mbH&Co	FRG
Enkel 11 kW	LAWEY	Lagerwey Windturbine	NL
Enkel 20 kW	LAWEY	Lagerwey Windturbine	NL
Enkel 35 kW	LAWEY	Lagerwey Windturbine	NL
ES 03-01	SCHUB	Schubert Elektrotechnik GmbH & Co KG	FRG
ES 10-11	SCHUB	Schubert Elektrotechnik GmbH & Co KG	FRG
ES 10-22	SCHUB	Schubert Elektrotechnik GmbH & Co KG	FRG
ES 1000 L	SCHUB	Schubert Elektrotechnik GmbH & Co KG	FRG
ES 20-100	SCHUB	Schubert Elektrotechnik GmbH & Co KG	FRG
ES 2000 L	SCHUB	Schubert Elektrotechnik GmbH & Co KG	FRG
ES3	ENSER	Energy Services Ltd.	UK
Esbjerg 1	ISVEST	I/S Vestkraft & Elsam	DK
F-19	CENEME	CENEME S.A.	E
F.H.W. 30	HUELL	Huellmann-Anlagenbau KG	FRG
Fasterholt 55kW	FAST	Fasterholt Maskinfabrik A/S	DK
Fasterholt 75kW	FAST	Fasterholt Maskinfabrik A/S	DK
Flair 8	VILLAS	Villas GmbH	AU
FLODA 600	VILLAS	Villas GmbH	AU
Folkec. 130 S	NORFOL	Nordvestjysk Folkecenter	DK
Folkecenter 75	NORFOL	Nordvestjysk Folkecenter	DK
Folkecenter13.5	NORFOL	Nordvestjysk Folkecenter	DK
G1000	BORNAY	Bornay	E
G200	BORNAY	Bornay	E
G5000	BORNAY	Bornay	E
GAMMA 60	AERIT	Aeritalia S.A.I. S.p.A.	I
Growian	MAN	MAN Technologie GmbH	FRG
GSS 10/6	GSS	GSS Power-Mills ApS	DK
HAWK 4.2 mtr.	TWST	T.W.Standivan Wind Turbine Co.	UK
HD 06512	WIKZ	Windkraft-Zentrale	FRG
HD 250 W	WIKZ	Windkraft-Zentrale	FRG
HD 312, HD 324	WIKZ	Windkraft-Zentrale	FRG
HE 1000 L	H-ENSY	H-Energiesystemen B.V.	NL
HE 1500 L	H-ENSY	H-Energiesystemen B.V.	NL
HE 1600	H-ENSY	H-Energiesystemen B.V.	NL
HE 2000	H-ENSY	H-Energiesystemen B.V.	NL
HE 3000	H-ENSY	H-Energiesystemen B.V.	NL
HE 600 L	H-ENSY	H-Energiesystemen B.V.	NL
HM-Rotor 20/56	HMOT	Heidelberg Motoren GmbH	FRG
HM-Rotor 20/60	HMOT	Heidelberg Motoren GmbH	FRG
HM-Rotor 300	HMOT	Heidelberg Motoren GmbH	FRG
hornet 25	TORNA	Tornado Wind Generators, Ltd.	UK
HSW-250	HSW	Husumer Schiffswerft	FRG
HSW-30	HSW	Husumer Schiffswerft	FRG
HVK 22/11	HVK	HVK	DK
HVK 22/12	HVK	HVK	DK
HVK 30/12	HVK	HVK	DK

B   List of WEC Types and WEC Manufacturers (contd.)

Type name	manufact. short name	manufacturer name	country
HVK 37/12	HVK	HVK	DK
HWP-330	HOWDEN	James Howden & Company	UK
HWP-60	HOWDEN	James Howden & Company	UK
HWP-750	HOWDEN	James Howden & Company	UK
Jacobs 1.8KW	JACO	Jacobs	IRL
Kano-Rotor 30kW	KAEHL	Kaehler-Maschinenbau GmbH	FRG
Kano-Rotor 600S	KAEHL	Kaehler-Maschinenbau GmbH	FRG
KEWT 1	GBO	GBO Energie-Systemen	NL
KONG 22/10	KONGST	Kongsted	DK
KRAM 15/8.3	KRAMS	Kramsbjerg-Mollen	DK
KURI 11/10.9	KURI	Kuriant-Maskinfabrik	DK
KURI 15/10	KURI	Kuriant-Maskinfabrik	DK
KURI 15/11	KURI	Kuriant-Maskinfabrik	DK
KURI 15/8	KURREY	Kuriantreymo	DK
KURI 18/10	KURI	Kuriant-Maskinfabrik	DK
LJ75A312	JELOU	Le Jeloux Sarl	F
LM 11	LM	LM Glasfiber A/S	DK
LM 7.75	LM	LM Glasfiber A/S	DK
LM 8.5	LM	LM Glasfiber A/S	DK
LMW 10/7	LMW	LMW Windenergy B.V.	NL
LMW 1000	LMW	LMW Windenergy B.V.	NL
LMW 1003	LMW	LMW Windenergy B.V.	NL
LMW 150	LMW	LMW Windenergy B.V.	NL
LMW 250	LMW	LMW Windenergy B.V.	NL
LMW 2500	LMW	LMW Windenergy B.V.	NL
LMW 3600	LMW	LMW Windenergy B.V.	NL
LMW 3600 (2,5kW	LMW	LMW Windenergy B.V.	NL
LMW 3600(2 blad	LMW	LMW Windenergy B.V.	NL
LMW 600	LMW	LMW Windenergy B.V.	NL
Lolland 80/18m	DAVID	A. Davidsen Nakskov A/S	DK
LS-1	WEGROU	Wind Energy Group Ltd.	UK
LW 11/35	LAWEY	Lagerwey Windturbine	NL
LW 15/75	LAWEY	Lagerwey Windturbine	NL
LW 24/10.6	LAWEY	Lagerwey Windturbine	NL
LW 45/15.6	LAWEY	Lagerwey Windturbine	NL
LW 5/5	LAWEY	Lagerwey Windturbine	NL
LW 50/15,6	LAWEY	Lagerwey Windturbine	NL
LW 55/15	LAWEY	Lagerwey Windturbine	NL
LW 60/15	LAWEY	Lagerwey Windturbine	NL
M 015-4	LUBING	Lubing Maschinenfabrik	FRG
M 100	MICON	Moerup Manufacturing Company, Micon A/S	DK
M 100/US	MICON	Moerup Manufacturing Company, Micon A/S	DK
M 108	WENERG	Wenergy DWT Industry A/S	DK
M 15 (25 kW)	RIVA	Riva Calzoni S.p.A.	I
M 15 (30 kW)	RIVA	Riva Calzoni S.p.A.	I
M 175kW	MICON	Moerup Manufacturing Company, Micon A/S	DK
M 200kW	MICON	Moerup Manufacturing Company, Micon A/S	DK
M 22	WINCON	Wincon Wind Systems Aps.	DK
M 30	RIVA	Riva Calzoni S.p.A.	I

B   List of WEC Types and WEC Manufacturers (contd.)

Type name	manufact. short name	manufacturer name	country
M 300 - 55 kW	MICON	Moerup Manufacturing Company, Micon A/S	DK
M 400kW	MICON	Moerup Manufacturing Company, Micon A/S	DK
M 450-150 kW	MICON	Moerup Manufacturing Company, Micon A/S	DK
M 450-250kW	MICON	Moerup Manufacturing Company, Micon A/S	DK
M 530-250kW	MICON	Moerup Manufacturing Company, Micon A/S	DK
M 55/11	MICON	Moerup Manufacturing Company, Micon A/S	DK
M 55/11	WINCON	Wincon Wind Systems Aps.	DK
M 55/16	WENERG	Wenergy DWT Industry A/S	DK
M 60/13/US	MICON	Moerup Manufacturing Company, Micon A/S	DK
M 66/13-US	WENERG	Wenergy DWT Industry A/S	DK
M 7	RIVA	Riva Calzoni S.p.A.	I
M 95	WENERG	Wenergy DWT Industry A/S	DK
M 95kW	MICON	Moerup Manufacturing Company, Micon A/S	DK
M110 (50Hz)	WINCON	Wincon Wind Systems Aps.	DK
M24 (250kW)	WENERG	Wenergy DWT Industry A/S	DK
M24 (350kW)	WENERG	Wenergy DWT Industry A/S	DK
M24 (400kW)	WENERG	Wenergy DWT Industry A/S	DK
M66/13	WINCON	Wincon Wind Systems Aps.	DK
Maglarp	STAENE	Statens Energiverk	S
MBB 15/30	MBB	Messerschmitt-Boelkow-Blohm GmbH	FRG
MK 015-6-3-1	LUBING	Lubing Maschinenfabrik	FRG
ML 015-6-3	LUBING	Lubing Maschinenfabrik	FRG
Model 2	AEROT	Aerotron Easter Achloa	UK
Model 3	AEROT	Aerotron Easter Achloa	UK
Monopteros 15	MBB	Messerschmitt-Boelkow-Blohm GmbH	FRG
Monopteros 50	MBB	Messerschmitt-Boelkow-Blohm GmbH	FRG
Monopterus 20	MBB	Messerschmitt-Boelkow-Blohm GmbH	FRG
Monopterus 30	MBB	Messerschmitt-Boelkow-Blohm GmbH	FRG
MP 5	RIVA	Riva Calzoni S.p.A.	I
MP5	HAWKER	Hawker Siddeley Power Eng. Ltd	UK
MS-1	WEGROU	Wind Energy Group Ltd.	UK
MS-2	WEGROU	Wind Energy Group Ltd.	UK
MS-2	WEGROU	Wind Energy Group Ltd.	UK
MS-3	WEGROU	Wind Energy Group Ltd.	UK
MS.2	WEGROU	Wind Energy Group Ltd.	UK
MSW-VAWT	KIDEN	Orientel Kiden CO.,LTD	J
MSW-VAWT	KIDEN	Orientel Kiden CO.,LTD	J
N 12/30	SUEWIN	Suedwind Windkraftanlagen	FRG
N 1245	SUEWIN	Suedwind Windkraftanlagen	FRG
N 715	SUEWIN	Suedwind Windkraftanlagen	FRG
N 718.5	SUEWIN	Suedwind Windkraftanlagen	FRG
Naesudden	STAENE	Statens Energiverk	S
NBK 100	BOHEM	Bohemen Energy Systems B.V.	NL
NBK 300	BOHEM	Bohemen Energy Systems B.V.	NL
NBK 600	BOHEM	Bohemen Energy Systems B.V.	NL
Newecs-25	FDO	Stork FDO WES B.V.	NL
Newecs-45	FDO	Stork FDO WES B.V.	NL
NIBE A	STATSM	Statsmoelle ("NIBE A/B")	DK
NIBE B	STATSM	Statsmoelle ("NIBE A/B")	DK

B List of WEC Types and WEC Manufacturers (contd.)

Type name	manufact. short name	manufacturer name	country
NN	AIR	Air Energy Systems	NL
Noah90kW	NOAH	Noah Energietechnik	FRG
NOAHGrundmodell	NOAH	Noah Energietechnik	FRG
Nordex 150 kW	NORDEX	Nordex A/S	DK
Nordex 150/26	NORDEX	Nordex A/S	DK
Nordex 225/250	NORDEX	Nordex A/S	DK
NTK - 130F	NOTANK	Nordtank Energy Group	DK
NTK - 75F	NOTANK	Nordtank Energy Group	DK
NTK 10/8	NOTANK	Nordtank Energy Group	DK
NTK 100/18.2	NOTANK	Nordtank Energy Group	DK
NTK 130/20.4	NOTANK	Nordtank Energy Group	DK
NTK 145 kW	NOTANK	Nordtank Energy Group	DK
NTK 150 XLR	NOTANK	Nordtank Energy Group	DK
NTK 22/10.5	NOTANK	Nordtank Energy Group	DK
NTK 22/10.8	NOTANK	Nordtank Energy Group	DK
NTK 22/11	NOTANK	Nordtank Energy Group	DK
NTK 30/11	NOTANK	Nordtank Energy Group	DK
NTK 300/31	NOTANK	Nordtank Energy Group	DK
NTK 400/35	NOTANK	Nordtank Energy Group	DK
NTK 45/15	NOTANK	Nordtank Energy Group	DK
NTK 45/16.9	NOTANK	Nordtank Energy Group	DK
NTK 450/37	NOTANK	Nordtank Energy Group	DK
NTK 55/11	NOTANK	Nordtank Energy Group	DK
NTK 55/14	NOTANK	Nordtank Energy Group	DK
NTK 55/15	NOTANK	Nordtank Energy Group	DK
NTK 55/16	NOTANK	Nordtank Energy Group	DK
NTK 65/14	NOTANK	Nordtank Energy Group	DK
NTK 65/16	NOTANK	Nordtank Energy Group	DK
NTK 75/16	NOTANK	Nordtank Energy Group	DK
NTK 99/22	NOTANK	Nordtank Energy Group	DK
NTK 150	NOTANK	Nordtank Energy Group	DK
ODD 18/10	ODDER	Odder-Mollen ApS	DK
ODD 18/11	ODDER	Odder-Mollen ApS	DK
Oestas 2 kW	OESTAS	Oestas	DK
OM-18.5kW	ODDER	Odder-Mollen ApS	DK
OR 30/12	OR.WIN	OR. Vindmoelle	DK
PEUI-10/2	GASE	Gas y electricidad	E
PEUI-10/3	GASE	Gas y electricidad	E
PG 10	AEE	Altern. Energie- und Elektriksyst. GmbH	FRG
Pionier II	POLYMA	Polymarin B.V.	NL
POUL 15/10.5	POULS	Poulsen	DK
POUL 30/11	POULS	Poulsen	DK
Poulsen II	SOEG	Soegrens Maskinfabrik	DK
PR 10/7	PR 10	PR 10	DK
Proven 1k	PROVEN	Proven Engeneering Prod. Ltd.	UK
Quadro	LAWEY	Lagerwey Windturbine	NL
REY 11/55	REYMO	Reymo DK A/S Maskinfabrik	DK
REY 15/8.5	REYMO	Reymo DK A/S Maskinfabrik	DK
REY 18.5/10.8	REYMO	Reymo DK A/S Maskinfabrik	DK

B   List of WEC Types and WEC Manufacturers (contd.)

Type name	manufact. short name	manufacturer name	country
REY 18/8.5	REYMO	Reymo DK A/S Maskinfabrik	DK
REY 20/95kW	REYMO	Reymo DK A/S Maskinfabrik	DK
REY 275	REYMO	Reymo DK A/S Maskinfabrik	DK
REY 3,5 kW	REYMO	Reymo DK A/S Maskinfabrik	DK
REY 30/150kW	REYMO	Reymo DK A/S Maskinfabrik	DK
REY 4/18.5	REYMO	Reymo DK A/S Maskinfabrik	DK
REY 55/15.6	REYMO	Reymo DK A/S Maskinfabrik	DK
RF 100	RATIER	Ratier-Figeac	F
RF 70	RATIER	Ratier-Figeac	F
Richborough	HOWDEN	James Howden & Company	UK
RIIS 17/10	RIISAG	Rijsager El-Vindmoller	DK
RIIS 22/10	RIISAG	Rijsager El-Vindmoller	DK
RIIS 35/14	RIISAG	Rijsager El-Vindmoller	DK
SAMS 80/18.7	SAMSOE	Samsoe Moellen	DK
Single	KAAL	Kaal - v.d. Linden	NL
sixmaster	LAWEY	Lagerwey Windturbine	NL
SJ 10/6.3	SJ WIN	SJ Windpower	DK
SJ 10/7	SJ WIN	SJ Windpower	DK
SM22	SONBJ	Sonebjerg Maskinfabrik	DK
SM30	SONBJ	Sonebjerg Maskinfabrik	DK
SM45	SONBJ	Sonebjerg Maskinfabrik	DK
SM55	SONBJ	Sonebjerg Maskinfabrik	DK
SME 22/10(V.2)	DANSME	Dansk Smedemesterforening	DK
SME 30/12	DANSME	Dansk Smedemesterforening	DK
SME 55/16	DANSME	Dansk Smedemesterforening	DK
SME 65/16	DANSME	Dansk Smedemesterforening	DK
SME 75/20	DANSME	Dansk Smedemesterforening	DK
SME 80/18	DANSME	Dansk Smedemesterforening	DK
SME 90/20	DANSME	Dansk Smedemesterforening	DK
SME 99/21.3	DANSME	Dansk Smedemesterforening	DK
Smedemoellen 80	DWTNAK	Danish Wind Turbines Nakskov	DK
SMEMOE 65/16.3	SMEMOE	Smedemoelle	DK
SMEMOE 80/18.7	SMEMOE	Smedemoelle	DK
SONE 22/10	SONBJ	Sonebjerg Maskinfabrik	DK
SONE 30/10	SONBJ	Sonebjerg Maskinfabrik	DK
SONE 45/12	SONBJ	Sonebjerg Maskinfabrik	DK
SONE 55/14	SONBJ	Sonebjerg Maskinfabrik	DK
SONE 55/15	SONBJ	Sonebjerg Maskinfabrik	DK
Svedana 80/50Hz	SVEDAN	Svedana A/S	DK
Svedana100/60Hz	SVEDAN	Svedana A/S	DK
SW-VAWT	KIDEN	Orientel Kiden CO.,LTD	J
SWEC2	ENRTCH	W. Schoenball Energietechnik	FRG
T 103	VIMOEL	K.M. Vindmoeller	DK
T 103 10/8	T 103	T 103	DK
T-17/100	DWT	Danish Wind Technology A/S	DK
T-17/80	DWT	Danish Wind Technology A/S	DK
Tellus/DWT 95	TELL	Tellus Energy Systems A/S	DK
TEMA 2	TEMA	TEMA s.p.a.	I
THG 5	SUEWIN	Suedwind Windkraftanlagen	FRG

B   List of WEC Types and WEC Manufacturers (contd.)

Type name	manufact. short name	manufacturer name	country
THY 11/8	THYMOE	Thy Moelle	DK
THY 7.5/7	THYMOE	Thy Moelle	DK
THY 7.5/7.5	THYMOE	Thy Moelle	DK
Trimble 10 KW	TRIM	Trimble	UK
TW 150	TACK	Tacke Windtechnik GmbH&Co KG	FRG
TW 250	TACK	Tacke Windtechnik GmbH&Co KG	FRG
TW 45	TACK	Tacke Windtechnik GmbH&Co KG	FRG
TW 500	TACK	Tacke Windtechnik GmbH&Co KG	FRG
TW 60	TACK	Tacke Windtechnik GmbH&Co KG	FRG
Twin	LAWEY	Lagerwey Windturbine	NL
Twinmaster	KAAL	Kaal - v.d. Linden	NL
TWS 125	TRASCO	Trasco	NL
TWS 175	TRASCO	Trasco	NL
UK T-1780	TELL	Tellus Energy Systems A/S	DK
UKEN 11/10	UKENDT	Ukendt	DK
UM70/10	AERWAT	Aerowatt International	F
UM70/2.5	AERWAT	Aerowatt International	F
UM70/5	AERWAT	Aerowatt International	F
V-S 130	VISYS	Vind-Syssel A/S	DK
V-S 150 (V.2)	VISYS	Vind-Syssel A/S	DK
V-S 180	VISYS	Vind-Syssel A/S	DK
V-S 225	VISYS	Vind-Syssel A/S	DK
V-S 270	VISYS	Vind-Syssel A/S	DK
V6	TURBO	Turbowinds, c/o TRECO S.A.	B
V86	TURBO	Turbowinds, c/o TRECO S.A.	B
Vanguard 95	WTI	W.T.I. (Ireland) Ltd.	IRL
Vanguard 95KW	WTI	W.T.I. (Ireland) Ltd.	IRL
VAWT 100 KW	VAWT	Vertical Axis Wind Turbines Ltd.	UK
VAWT 1200kW	VAWT	Vertical Axis Wind Turbines Ltd.	UK
VAWT 130 KW	VAWT	Vertical Axis Wind Turbines Ltd.	UK
VAWT 850/500kW	VAWT	Vertical Axis Wind Turbines Ltd.	UK
VB-11	IRAS	Maskinfabrikken IRAS,Esbjerg	DK
VEND 24/12	VENDEL	Vendelbo	DK
Vertax DD	MMEW	MMEW Marketing Ltd.	UK
Vertax HD	MMEW	MMEW Marketing Ltd.	UK
VEST 100/20	VESTAS	Vestas Danish Wind Technology A/S	DK
VEST 150/25	VESTAS	Vestas Danish Wind Technology A/S	DK
VEST 200/25	VESTAS	Vestas Danish Wind Technology A/S	DK
VEST 22/10	VEST/H	Vestas/HVK	DK
VEST 225/27	VESTAS	Vestas Danish Wind Technology A/S	DK
VEST 250/27	VESTAS	Vestas Danish Wind Technology A/S	DK
VEST 30/10	VESTAS	Vestas Danish Wind Technology A/S	DK
VEST 30/11	VEST/H	Vestas/HVK	DK
VEST 30/18	VESTAS	Vestas Danish Wind Technology A/S	DK
VEST 55/15	VEST/H	Vestas/HVK	DK
VEST 55/16	VESTAS	Vestas Danish Wind Technology A/S	DK
VEST 55/16(V.2)	VESTAS	Vestas Danish Wind Technology A/S	DK
VEST 55/17	VESTAS	Vestas Danish Wind Technology A/S	DK
VEST 75/15	VESTAS	Vestas Danish Wind Technology A/S	DK

Type name	manufact. short name	manufacturer name	country
VEST 75/17(V.2)	VESTAS	Vestas Danish Wind Technology A/S	DK
VEST 90/18	VESTAS	Vestas Danish Wind Technology A/S	DK
VEST 90/20	VESTAS	Vestas Danish Wind Technology A/S	DK
VEST 99/20	VESTAS	Vestas Danish Wind Technology A/S	DK
VIBL 7/6	VIBLIS	Vind Blis	DK
VIMOE 10/7	VIMOEL	K.M. Vindmoeller	DK
VIMOE 10/7.5	VIMOEL	K.M. Vindmoeller	DK
VIMOE 4/5.4	VIMOEL	K.M. Vindmoeller	DK
Vortex 1000	NATPOW	Natural Power Systems, Inc.	UK
Vortex 200	NATPOW	Natural Power Systems, Inc.	UK
Vortex 6000	NATPOW	Natural Power Systems, Inc.	UK
W 100	SOVENT	Solavent Dr.Ing. Theo Bracke	FRG
W 150 XT	WINCON	Wincon Wind Systems Aps.	DK
W 150/25m	WIWOR	Wind World A/S	DK
W 150/27m	WIWOR	Wind World A/S	DK
W 200	WINCON	Wincon Wind Systems Aps.	DK
W 22	WINCON	Wincon Wind Systems Aps.	DK
W 3	WINCON	Wincon Wind Systems Aps.	DK
W 300	SOVENT	Solavent Dr.Ing. Theo Bracke	FRG
W 300/25.4	WINCON	Wincon Wind Systems Aps.	DK
W 400/26	WINCON	Wincon Wind Systems Aps.	DK
W-1960	WIWOR	Wind World A/S	DK
W-2250 (50Hz)	WIWOR	Wind World A/S	DK
W-2300	WIWOR	Wind World A/S	DK
W-2500 220kW	WIWOR	Wind World A/S	DK
W-2800 150kW	WIWOR	Wind World A/S	DK
W110/XT (50Hz)	WINCON	Wincon Wind Systems Aps.	DK
WE 03-03	SCHUB	Schubert Elektrotechnik GmbH & Co KG	FRG
WG 12	GBO	GBO Energie-Systemen	NL
WG 16	GBO	GBO Energie-Systemen	NL
WG 910	HARB	A. Harbarth	FRG
WIMA 11/8	WINMAT	Wind-Matic A/S	DK
WIMA 130/20	WINMAT	Wind-Matic A/S	DK
WIMA 200	WINMAT	Wind-Matic A/S	DK
WIMA 22/10	WINMAT	Wind-Matic A/S	DK
WIMA 22/32	WINMAT	Wind-Matic A/S	DK
WIMA 30/10	WINMAT	Wind-Matic A/S	DK
WIMA 30/12	WINMAT	Wind-Matic A/S	DK
WIMA 45/12	WINMAT	Wind-Matic A/S	DK
WIMA 45/14	WINMAT	Wind-Matic A/S	DK
WIMA 55/14	WINMAT	Wind-Matic A/S	DK
WIMA 55/15	WINMAT	Wind-Matic A/S	DK
WIMA 75/15	WINMAT	Wind-Matic A/S	DK
WIMA 75/17(V.2)	WINMAT	Wind-Matic A/S	DK
WIMA 99/19	WINMAT	Wind-Matic A/S	DK
WINC 198/21	WINCON	Wincon Wind Systems Aps.	DK
WINC 95/19.6	WINCON	Wincon Wind Systems Aps.	DK
WINC 99/20	WINCON	Wincon Wind Systems Aps.	DK
WINC 99/21	WINCON	Wincon Wind Systems Aps.	DK

B   List of WEC Types and WEC Manufacturers (contd.)

Type name	manufact. short name	manufacturer name	country
Wind-World 1	WIWOR	Wind World A/S	DK
Wind-World 120	WIWOR	Wind World A/S	DK
Wind-World 150	WIWOR	Wind World A/S	DK
Wind-World 160	WIWOR	Wind World A/S	DK
Wind-World 180	WIWOR	Wind World A/S	DK
Wind-World 2	WIWOR	Wind World A/S	DK
Wind-World 95	WIWOR	Wind World A/S	DK
WINDANE 10/18	DWT	Danish Wind Technology A/S	DK
WINDANE 12/18	DWT	Danish Wind Technology A/S	DK
WINDANE 12/20	DWT	Danish Wind Technology A/S	DK
WINDANE 12/25	DWT	Danish Wind Technology A/S	DK
WINDANE 12/30	DWT	Danish Wind Technology A/S	DK
WINDANE 12/40	DWT	Danish Wind Technology A/S	DK
WINDANE 12/50	DWT	Danish Wind Technology A/S	DK
WINDANE 17/80	DWT	Danish Wind Technology A/S	DK
WINDANE 19/130	DWT	Danish Wind Technology A/S	DK
WINDANE 19/95	DWT	Danish Wind Technology A/S	DK
WINDANE 19/95	DWT	Danish Wind Technology A/S	DK
WINDANE 31/300	DWT	Danish Wind Technology A/S	DK
WINDANE 34/400	DWT	Danish Wind Technology A/S	DK
WINDANE 40/750	DWT	Danish Wind Technology A/S	DK
WINDANE 5/4	DWT	Danish Wind Technology A/S	DK
WINDANE 9/11	DWT	Danish Wind Technology A/S	DK
Windgen.2000W	ELEKTR	ELEKTRO GMBH	CH
Windgen.50 W	ELEKTR	ELEKTRO GMBH	CH
Windgen.500W	ELEKTR	ELEKTRO GMBH	CH
Windgen.6000W	ELEKTR	ELEKTRO GMBH	CH
Windiesel	LAWSON	Sir Henry Lawson-Tancred	UK
WindMaster 100	HMZ	HMZ Belgium N.V.	B
WindMaster 150	HMZ	HMZ Belgium N.V.	B
WindMaster 175	HMZ	HMZ Belgium N.V.	B
WindMaster 200	HMZ	HMZ Belgium N.V.	B
WindMaster 225	HMZ	HMZ Belgium N.V.	B
WindMaster 250	HMZ	HMZ Belgium N.V.	B
WindMaster 300	HMZ	HMZ Belgium N.V.	B
WindMaster 500	HMZ	HMZ Belgium N.V.	B
Windpaq	BERWOU	Berewoud Energie B.V.	NL
Windvang 125.40	BERWOU	Berewoud Energie B.V.	NL
Windvang 160.6	BERWOU	Berewoud Energie B.V.	NL
Windvang 200.15	BERWOU	Berewoud Energie B.V.	NL
Windvang 65.8	BERWOU	Berewoud Energie B.V.	NL
Windvang 80.15	BERWOU	Berewoud Energie B.V.	NL
WKA 60	MAN	MAN Technologie GmbH	FRG
WL40LC	GRYLLS	Grylls Windtech	UK
wm 15 S	WINMAT	Wind-Matic A/S	DK
wm 17 s	WINMAT	Wind-Matic A/S	DK
WM 19S	WINMAT	Wind-Matic A/S	DK
WM 22S	WINMAT	Wind-Matic A/S	DK
WPI	WIND	Windpower & Co. Ltd.	UK

B   List of WEC Types and WEC Manufacturers   (contd.)

Type name	manufact. short name	manufacturer name	country
WPN-ZD 0.2	HAZSOL	Hazsolar Energias Alternativas	E
WPN-ZD 0.8	HAZSOL	Hazsolar Energias Alternativas	E
WPS 10	POLENK	Polenko B.V.	NL
WPS 20	POLENK	Polenko B.V.	NL
WPS-35-550 kW	HOLEC	Holec Projects BV	NL
WPS10A15	POLENK	Polenko B.V.	NL
WPS10SM15	POLENK	Polenko B.V.	NL
WPS11A40	POLENK	Polenko B.V.	NL
WPS16A60	POLENK	Polenko B.V.	NL
WPS16SM60	POLENK	Polenko B.V.	NL
WPS18A100	POLENK	Polenko B.V.	NL
WPS18SM100	POLENK	Polenko B.V.	NL
WPS30/3	POLENK	Polenko B.V.	NL
WPS30/4	POLENK	Polenko B.V.	NL
WPS30SM300	POLENK	Polenko B.V.	NL
WPS82M10	POLENK	Polenko B.V.	NL
WPS8A10	POLENK	Polenko B.V.	NL
WR 100	PIBAM	Pintsch Bamag	FRG
WR 150	TACK	Tacke Windtechnik GmbH&Co KG	FRG
WR 65/17	BSW	Burbacher Stahl-und Waggonbau	FRG
WR375kW	BSW	Burbacher Stahl-und Waggonbau	FRG
WR80/18	BSW	Burbacher Stahl-und Waggonbau	FRG
WTS 75-3	BOVING	Boving-KMW Turbin AB	S
WTW 7	BCA	B.C.A.	NL
WV15	ELEKT	Elektro	NL
WV35	ELEKT	Elektro	NL
WV50	ELEKT	Elektro	NL
Y12	AIOLOS	Aiolos Windmolengroep B.V.	NL

C   Questionnaires Used for EUROWIN Database

Fraunhofer-Institut for Solar Energy Systems - Freiburg - FRG
single machine / (windpark)

date   : _____
country   : _____
plant nr.   : _____
year   : _____
month   : _____

OPERATOR QUESTIONNAIRE FOR THE ESTABLISHMENT OF A EUROPEAN WIND TURBINE DATA BASE

operator
operator name   : _____
contact person : _____
street   : _____
Po-Box   : _____
region   : _____
town   : _____
tel.   : _____
fax   : _____

windturbine
manufacturer : _____
type name   : _____
rated power   : _____ [kW]
diameter   : _____ [m]
swept aerea   : _____ [m²]
hub height   : _____ [m]

OPERATIONAL WINDTURBINE

use : O grid connected O private   O public
           O stand allone   O industrial O utility

location : - coordinates X : _____   altitude : _____ [m]
                               Y : _____
                - postal code   : _____
                - name of site   : _____ :

local data : - surface roughness class   : _____
                  - mean annual wind speed   : _____ [m/sec]
                     measured at :   O 10 m
                                      O hub height
                                      O other, specify : _____ [m]
                                      O next metereological station
                  - Weibull parameters   scale parameter C : _____
                                         shape parameter K : _____

date of electrical connection : : y/19___ / ; m/____ / ; d/____ /

costs :   national currency : _____
              machine ex works   : _____
              foundation   : _____
              transport   : _____
              erection   : _____
              elect.connection   : _____
              consulting   : _____
              land costs   : _____

operational data: energy production, total : _____ [kWh]
                      operational time   : _____ [h]
                      downtime   : _____ [h]

Additional:
please include any additional information regarding repairs, changes relative to this windturbine
: _____

Fraunhofer-Institut for Solar Energy Systems - Freiburg - FRG

single machine / (windpark)

date	: _____
country	: _____
plant nr.	: _____
year	: _____
month	:

MONTHLY QUESTIONNAIRE FOR THE ESTABLISHMENT OF A EUROPEAN WIND TURBINE DATA BASE

operator name : _____   type name : _____

location   : _____   rated power : _____ [kW]

MONTHLY OPERATIONAL DATA

energy :   total production   : _____ [kWh]
            used from grid   : _____ [kWh]
            sold to grid,day   : _____ [kWh]
            sold to grid,night   : _____ [kWh]

operational time   : _____ [h]

mean monthly wind speed 10m   : _____ [m/sec] height : _____ [m] : _____ [m/sec]
at site   hub height : _____ [m/sec]

FAILURES EXPERIENCED

component :	subcomponent	downtime [h]	repair (R) exchange(E) tot.breakdown (T) service (S)	reason (see below)	costs [nat.curr.]: _____
O entire unit		_____	_____	_____	_____
O not specified		_____	_____	_____	_____
O rotor	hub	_____	_____	_____	_____
	blades	_____	_____	_____	_____
	airbrakes	_____	_____	_____	_____
O drive train	shaft	_____	_____	_____	_____
	bearings	_____	_____	_____	_____
O gearbox		_____	_____	_____	_____
O clutch		_____	_____	_____	_____
O brake		_____	_____	_____	_____
O generators		_____	_____	_____	_____
O yaw-system	motor	_____	_____	_____	_____
	mechanics	_____	_____	_____	_____
O tower		_____	_____	_____	_____
O foundation		_____	_____	_____	_____
O grid connection		_____	_____	_____	_____
O control-system	electronic	_____	_____	_____	_____
	mechanic	_____	_____	_____	_____
O nacelle	platform	_____	_____	_____	_____
	subcomp.				

---

reason

A lightning	D crack	G short circuit	K other
B wear	E stop, high rotational speed	H stop, high wind speed	
C fatigue fracture	F fail in construction	I overheating	

Fraunhofer-Institut for Solar Energy Systems - Freiburg - FRG   page : 1

(single machine) / windpark

date ..
country ..
plant nr. ..
year ..
month ..

MONTHLY QUESTIONNAIRE FOR THE ESTABLISHMENT OF A EUROPEAN WIND TURBINE DATA BASE

operator name ..
name of windpark ..
location ..

no. of identical units : ..
type name : ..
rated power : .. [kW]
hub height : .. [m]

mean monthly wind speed at site   10m : _____ [m/sec] ; hub height : _____ [m/sec] ; height : _____ [m] .. : _____ [m/sec]

MONTHLY OPERATIONAL DATA

No.	ENERGY GENERATION		OPERAT. TIME	DOWN TIME	FAILURES EXPERIENCED codes for component, extent and reason see page 4		
	total [kWh]	used.grid [kWh]	[h]	[h]	component /- sub-comp.	extent / reason	costs [nat.curr.]:
01					/	/	
02							
03							
04							
05							
06							
07							
08							
09							
010							

CODES FOR FAILURES EXPERIENCED

component / subcomp.	component / sub-comp.	component / sub-comp.	component / sub-comp.
1 entire unit : ..	4 drive train : a)shaft	8 generators : a)motor	12 grid connection : ..
2 not specified	b)bearings	9 yaw-system : b)mechanic	13 control system : a)electric
3 rotor : a)hub	5 gearbox ..		b)mechanic
b)blades	6 clutch ..	10 tower ..	14 nacelle : a)platform
c)airbrakes	7 brake ..	11 foundation ..	b)sub-comp.

extent R repair
      E exchange
      T total brake down
      S service

reason A lightning   D crack   G short circuit   K other
      B wear   E stop, high rotational speed   H stop, high wind speed
      C fatigue fracture   F fail in construction   I overheating

ADDITIONAL: _____

_____

_____

_____

_____

_____

_____

_____